OTHER BOOKS BY THE AUTHOR

FICTION

Aleutian Fury
Operation GB
Best Time of Year
Chelydra Serpentina
The Sun That Did Not Rise

NONFICTION

PC Patrol Craft of World War II
Seabag of Memories
How to Save Energy and Money at Home and on the Highway
Energy – Its Mystique, Uses, and Consequences
Golf is Like Love

OTHER

A Boy From Queens
My Navy Days
The Crimson Vortex – And Other Stories, Vignettes, and Essays

FUNNY FACTS OF PHYSICS

Wm. J. Veigele, Ph. D.

FUNNY FACTS OF PHYSICS
by
William J. Veigele, Ph. D.

Astral Publishing Co.

First Edition

No part of this book may be reproduced or transmitted in any form or by any means, graphic, electronic or mechanical, including photocopying, recording, taping or by any information storage or retrieval system, without permission in writing from the author or publisher.

Copyright © 2017: Wm. J. Veigele
Create Space
ISBN-13: 978-1979505673
ISBN-10: 1979505675

Library of Congress Cataloging in Publication Data
William J. Veigele,
www.astralpublishing.com

Funny Facts of Physics
paper, 218 pages

1. Nonfiction, 2. Physics 3. Gravity, 4. Relativity, 5. Quantum Theory, 6.Cosmology, 7. Metaphysics 8. Pseudophysics

Printed in the United States of America

DEDICATION

I dedicate this book to YOU . . .
the person with an inquisitive
mind for whom I wrote this book .
. . . YOU . . . the reader.

NOTE

In the text I use the word human. In each case I mean only the sapien species of the homo (human) genus, the homo sapien.

TABLE OF CONTENTS

PREFACE. 17
PART ONE – EVERYDAY– CLASSICAL PHYSICS. 21
1. In the Middle. 23
2. Why Fish Don't Freeze. 33
3. Nothing is Black. 39
4. Energy Does Not Exist. 43
5. A Nickel's Worth of Energy. 49
6. You Cannot Go in Reverse 51
PART TWO – GRAVITY. 57
7. How to Lose Weight. 59
8. The Moon is Falling. 63
9. For a Good Stretch Try a Black Hole. 69
PART THREE – RELATIVITY. 73
10. The Fastest. 75
11. Faster Than the Fastest. 77
12. Shrink Happens. 81
13. How to Stay Young. 85
14. How to Bulk Up. 89
PART FOUR – QUANTUM THEORY. 91
15. Particle or Wave. 93
16. Schrödinger's Patchwork. 99
17. You Can Never Be Certain. 103
18. Some Actions are Spooky. 107
19. Bathed in Photons. 111

TABLE OF CONTENTS (Cont'd.)

20. Exciting Life of a Photon. 115
21. Colder Than the Coldest. 119
22. The Virtual Vacuum. 123
PART FIVE – COSMOLOGY. 125
23. I See a Star 127
24. Time Will Tell. 131
25. No Time Before Time. 137
26. No Space Outside of Space. 139
27. It's All In the Dark. 141
28. It Must Be Jelly. 145
29. 137 – The Universe's PIN Code. 149
PART SIX – FUTURE PHYSICS. 153
30. The End of Physics. 155
31. Tie it All Together. 159
32. Physics Has Many Futures. 163
EPILOGUE. 167
Epilogue. 169
APPENDICES. 177
1. Preface. Log Scale and Scientific Notation. 179
2. Ch. 4. Energy Does Not Exist. 181
3. Ch. 7. How to Lose Weight. 183
4. Ch. 8. The Moon is Falling. 187
5. Ch. 9. For a Good Stretch Try a Black Hole. 189
6. Ch. 10. The Fastest. 191
7. Ch. 12. Shrink Happens. 193

TABLE OF CONTENTS (Cont'd.)

8. Ch. 13 How to Stay Young.........................195
9. Ch. 14. How to Bulk Up. 197
10. Ch. 16. Schrödinger's Patchwork..................199
11. Ch. 17. You Can Never Be Certain................201
12. Ch. 19. Bathed in Photons........................203
13. Ch. 21. Colder Than the Coldest..................205
14. Ch. 22. The Virtual Vacuum......................209
15. Ch. 24. Time Will Tell...........................211
INDEX..213
ABOUT THE AUTHOR................................217

LIST OF FIGURES

Figure 1 – 1. Mass Size Relationships 24
Figure 1 – 2. Electromagnetic Spectrum In Wavelengths. 26
Figure 2 – 1. Liquid Volume versus Temperature. 34
Figure 2 – 2. Typical Liquid Density versus Temperature. 34
Figure 2 – 3. Water Volume versus Temperature. 35
Figure 2 – 4. Water Density versus Temperature. 36
Figure 2 – 5. Bond Angle in Water. 37
Figure 3 – 1. Umbra and Penumbra. 41
Figure 3 – 2. Earth, Moon, Sun with Eclipse. 41
Figure 4 – 1. The Energy Ladder. 46
Figure 8 – 1. Tangent to a Radius of a Circle. 65
Figure 8 – 2. Trajectories of a Ball and the Moon. 66
Figure 9 – 1. Spaghettification. 70
Figure 11 – 1. Cerenkov Radiation. 78
Figure 12 – 1. Relativistic Length Contraction of a Car. 81
Figure 12 – 2. Relativistic Length Contraction of a Train. 82
Figure 12 – 3. Relativistic Contraction of a Baseball. 82
Figure 12 – 4. Graph of γ vs. Speed Relative to c. 83
Figure 15 – 1. Single Slit Pattern for Light. 94
Figure 15 – 2. Double Slit Pattern for Light 95
Figure 20 – 1. The Andromeda Galaxy 115
Figure 21 – 1. Comparison of Temperature Scales. 121
Figure 24 – 1. Earth's Orbit and Moon's Orbit. 134
Figure 28 – 1. Gravitational Waves. 145
Figure 31 – 1. Sketch of Vibrating Strings. 159
Figure A7 – 1. Weight Inside and Outside of the Earth. 182
Figure A7 – 2. Composition of the Earth. 182
Figure A21 – 1. Energy Level Diagram. 204
Figure A21 – 2. Inverted Energy Level Diagram. 205
Figure A22 – 1. B B Irradiance vs. λ and T. 208
Figure A22 – 2. Cosmic Background Radiation vs. λ. 208

LIST OF TABLES

Table 1 – 1. Ranges of Physical Characteristics. 23
Table 1 – 2. Sound Frequency Ranges For Animals. 27
Table 1 – 3. Sound Power Levels From Various Sources. . . . 27
Table 8 – 1. Horizontal Distance versus Heights and Speeds. 64
Table 8 – 2. Speeds to Orbit Earth at Different Heights. 67
Table 12 – 1. Relativistic Length Contraction versus Speed. . 84
Table 13 – 1. Time Dilation at Relativistic Speeds. 86
Table 14 – 1. Ratio of a Mass at Speed v to its Rest Mass. . . 90
Table A19 – 1. Relative Permittivity and Permeability. 201

Funny Facts of Physics

ABOUT REFERENCES

In this book I am following a trend away from tradition. Usually in books, all or most factual statements are awarded a reference to the source, the origin of the work, and to the authors or researchers. This is usually done by citing a journal article, book, or oral presentation with the names of the contributors to those media. Following that tradition this book could have a reference section with hundreds of entries. They would be unnecessary for the reader.

Research in physics is always built on earlier work. How far back should I go in citing work to finally cite the original work? Where should one draw the line in citing one source but not another earlier one? In chapters involving mechanics should I go way back and reference Sir Isaac Newton's *Philosophiæ Naturalis Principia Mathematica* of 1687?

I believe this procedure of including references in a book is no longer required and is archaic. This reasoning is because all the information given in this book (and most of what is of interest in science and other disciplines) is available on the internet without the old necessity of searching through library archives and finding appropriate volumes on shelves and sitting at a table and pouring through them. All that effort and time consumed can be reduced to sitting at a computer and accessing the research of interest in much less time. Original research is the content of the book, but its references do not appear here. What is original in this book is my combination of words, my style, and my prose in describing the facts that are readily available.

With respect to physics writing, it tends to be duplicative, which to some persons may appear as plagiarism. This duplication occurs because physics uses precise expressions and

Funny Facts of Physics

exact definitions. These do not allow for stating facts or laws in a variety of ways. What one person writes for a law is what any other person would write for that law. In physics words have precise and singular meanings. In everyday language or literature it is different. Words can have more than one meaning, or more than one word can be used for the same idea. For example, the word "momentum" in a thesaurus and a dictionary is defined as impulse, force, forcefulness, strength, impetus, way, and speed. Neither of these definitions is the one and only one used in physics, I. e., momentum equals mass times velocity. In physics, momentum is defined as that and only that regardless of who uses it, when it is used, and in what context it is used.

My lack of citations is not that I don't respect and give credit to all persons who have contributed to the body of knowledge and theories of physics. I certainly do respect their work. I have used much of it in my own research, and other physicists have built on my work in their research. This is how physics progresses.

I encourage readers, however, if they are so inclined, to search out the sources of the physics included in this book on the internet or elsewhere and read first hand accounts of these great accomplishments.

PREFACE

Humans live sort of in the middle of everything that exists in the universe, from the largest to the smallest scales. I don't mean that they are in the exact center but somewhere near the middle with respect to mass, size, strength, speed, lifetimes, and the range of their senses such as sight and hearing. By being in the middle I mean somewhere in the middle on a factors-of-ten scale. See Appendix P. This middle position limits our ability to directly observe the largest and the smallest parts of the universe. This restricted view conditions humans to not be able to see directly the entire universe as it really is. As we learn more about the universe, therefor, much of what we learn seems strange.

These limitations mean that humans have a restricted experience of all that the universe offers. We experience the universe from the middle of its structure, and have no direct observation of the very large sizes, the very small sizes, the very large masses, the very tiny masses, and the very fast speeds. Humans classify and understand the world from their limited ability to sense it, and tend to think that what they experience is the way the universe is and works even at the extreme scales that they cannot directly experience. On the extreme scales, however, in many instances the world is different from what we expect. At the extremes it might even look funny.

Our Earth is very diverse with flora, fauna, and the inanimate materials all about us. It is a world of immense variety, and it's difficult to think that humans can consider it as one whole entity called nature. Then there are the worlds beyond Earth of stars and galaxies and much more. It's almost inconceivable that humans can understand and explain all this. But we can. Or at least we have explained much of it and are progressing toward further explanations. We humans may be dwarfed by the size of

Funny Facts of Physics

the universe, but we are giants in our ability to comprehend it.

To achieve that goal of understanding and explaining nature and the universe, we employ science which is a system, a paradigm. By that I mean we first simplify our picture of the world and assume ideal conditions. When we understand that simple picture we then add in its complications and work to understand their effects on the simpler picture.

To start we define the basic properties common to everything. We define nature in terms of three fundamental Dimensions: Mass, Length, and Time. Science is an attempt to understand everything about us and explain it in terms of these three fundamental Dimensions. Scientists observe natural objects, events, and phenomena and build conceptual models that are speculative replicas of what they see. In physics these models are translated into mathematical models using equations in which the terms in the equations represent space, time, objects and their properties, and the states of being and becoming of those objects. These mathematical models hopefully can be used not only to explain what was observed but to predict changes in them and even predict new objects, events, and phenomena. Physicists then do experiments to confirm the predictions of the models. When experiments agree with the models' predictions and succeed in explaining observations the models are called theories. A theory then is a successful explanation, including a mathematical model, of observations. And all observations are facts.

So physics deals with facts. And physics is fun. But at times some of the facts of physics seem funny. They are not funny in the sense of being laughable, only some times they seem strange. That's because some facts about the world, especially at the extreme large or small scales, differ from our preconceived ideas of or our familiar experiences in our everyday world because we live in the middle.

You may be surprised by the ideas included in this book that involve facts I call funny. Some of them you may be familiar

Preface

with but have misinterpreted, and some you may be unfamiliar with. So now you'll see some facts differently and some maybe for the first time. Some of the facts I'll present may be new to you. In each case I have tried to explain away their funny or peculiar or unexpected nature so they become accepted, rational, and familiar.

Of course this book is not a text or compendium of physics. There is much more to physics than what I have explored here, especially recent discoveries and developments. I have simply selected a few items from some obvious areas of physics for illustration. The physics presented here is not complete or always as rigorous as it could be, but hopefully it is not wrong.

I divided the book into six sections proceeding from the more familiar world of Everyday which is explained by Classical Physics to the increasingly less familiar worlds of Gravity, Relativity, Quantum Theory, Cosmology, and to what may be Future Physics.

You, the reader to whom this book is dedicated, need not worry about long discussions and a lot of special material to read. Each chapter is short and to the point, and the chapters stand alone. You can read them in any order. You won't have to worry about a lot of physics or mathematics or think you should know a bunch of equations you should be able to solve. You can read this book without needing scientific or mathematical knowledge. In this book I explain the physical ideas and facts I have selected. I use as little mathematics as possible, and when some appears I explain it. Also I include Appendices, where I explain the physics and mathematics further, for those who are inclined to read them.

This book is written for you, so in most cases I present it in the second person singular–you–. So settle back, read on, and enjoy some of the funny facts of physics.

Funny Facts of Physics

PART ONE

EVERYDAY – CLASSICAL PHYSICS

Funny Facts of Physics

In The Middle

CHAPTER 1

IN THE MIDDLE

Humans are considered the only animals that can apply reason, use logic, recall the past, and plan for the future. They alone possess emotions and can construct and appreciate art and science. Humans build and destroy and engage in sports and warfare. Most religions propose that everything in the universe was made for humankind and that they are the ultimate objects of creation, made in God's image. Scientists and philosophers have proposed an Anthropic Cosmological Principle one version of which states that the creation of the universe was fine tuned so that humans could appear. Everything in the universe was made so they could exist. Humans are the ultimate beings of creation. If humans are the ultimate objects in the universe, a funny fact of physics is that humans are not the biggest, smallest, fastest, or the most capable of all things that exist. They occupy a middle role in the universe as shown in Figure 1 – 1.

This recognition of being sort of in a middle position with respect to all things in the universe can also be summarized in Table 1 – 1 for orders of magnitudes including times scales.

	Mass (kgm)	Size (m)	Lifetime (sec)
Universe	10^{53}	10^{26}	10^{20}
Human	10^{2}	1	10^{9}
Atoms	10^{-31}	10^{-15}	10^{-32}

Table 1 – 1. Ranges of Physical Characteristics

Funny Facts of Physics

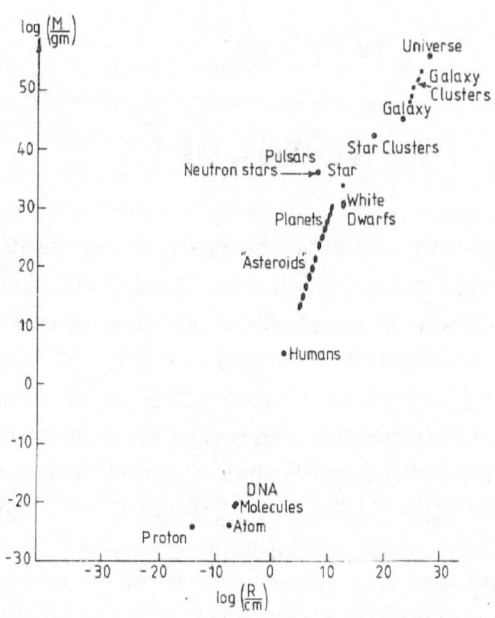

Figure 1 – 1. Mass Size Relationships

Because of this middle position, you have a limited range of your physical abilities and senses. Those limitations prevent you from directly experiencing the entire universe at both the very large and the very small scales. For illustration, Table 1 – 1 shows examples of the range of human physical characteristics of size, mass, and lifetime compared to the equivalent characteristics of the large and small features of the universe. On the logarithm scale human values fit somewhere near the middle of the range of values. As shown in Figure 1 – 1 and Table 1 – 1, humans experience masses, dimensions, and times that are not nearly as large or as small as those that exist in the universe from the microscopic to the cosmological ranges. They do not have an intimate sense of the very large and the very small. In many ways

In The Middle

humans are in the middle of the ranges of objects in the universe.

Similarly, humans are in the middle of the scale of physical abilities of all living things. Given below are some examples.

Strength

Ants can carry as much as 5,000 times their body weight. Elephants can lift with their trunks only about 5 % of their weight or 600 pounds and can carry on their backs only 25 % of their weight or about 3,000 pounds. By contrast humans can lift, on the average, their body weight. The largest weight lifted unaided by a man is 1,015 pounds, about five times his body weight. Humans can carry on their backs 50 pounds, about one third of their body weight. So they fit in the middle between ants (among the smallest creatures) and elephants (among the largest creatures) with respect to strength.

Agility

We are not as agile as house flies, humming birds, bats, or monkeys but more agile than hippopotami and walruses. Humans are somewhere in between various species with respect to agility.

Speed

Turtles and tortoise move at agonizingly slow speeds. The fastest creature is the peregrine falcon that can reach 200 miles per hour in a dive. The fastest land animal is the cheetah which can dash at 60 miles per hour. By comparison the fastest average speed recorded for a human over 200 m is 22.36 miles per hour. The fastest sprint speed over 10 m was 26.9 mph. So humans fit in somewhere in the middle of speeds of various living things.

As for the ranges of humans' senses, as compared to other living creatures, here are some examples.

Range of vision and the Electromagnetic Spectrum

The electromagnetic spectrum shown in Figure 1 – 2 ranges in wavelengths from 10^4 meters to 10^{-12} meters, from radio waves through visible light and onto gamma rays. That range covers sixteen orders of magnitude (multiples of ten). Also, shown are the sizes of the wavelengths of those frequencies compared to other objects. For example microwaves have a wavelength of about the height of a person while gamma rays have a wavelength of about the size of the nucleus of an atom.

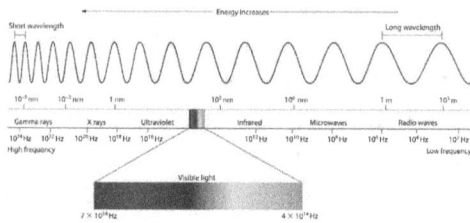

Figure 1 – 2. Electromagnetic Spectrum in Wavelengths

The portion of the electromagnetic spectrum you see is the visible section including red, orange, yellow, green, blue, indigo, and violet with wavelengths around 10^{-5} meters. That limited range is all you can see with your eyes. Just think what you are missing not seeing at least infrared and ultraviolet like some other creatures can.

This visible range of the electromagnetic spectrum is a very small fraction of the total spectrum, and the wavelength 10^{-5} m is near the logarithm middle of the spectrum. So with respect to experiencing the universe with your eyes your ability is limited to a small segment of the middle range of all that could be possible.

Range of Hearing

Sounds that humans can hear span the acoustic frequency range from about 20 – 20,000 Hz (Hertz is the name for the unit cycles per second). Human ears are most sensitive to sounds in the 2000 – 5000 Hz range which includes music and conversation. Responses to lower than 20 Hz are not heard but may be sensed as vibrations. It was said that Adolph Hitler arranged speaker systems around his rallies that broadcast an acoustic signal at 7 Hz. His advisors thought this acoustic signal would penetrate people's bodies and stimulate their emotions.

In The Middle

Acoustic signals greater than 20,000 Hz are not heard but sensed, possibly as pain. That most sensitive range of hearing for humans, 2000 - 5000 Hz, is about in the middle of the entire range on a logarithm scale. So humans can hear only a small range of acoustic vibrations that is in the middle of the entire possible range. Comparisons of frequency ranges for humans and other creatures are shown in Table 1 – 2.

Animal	Hearing range in Hertz
Humans	20 – 20,000
Bats	2000 – 110,000
Elephant	16 – 12,000
Fur Seal	800 – 50,000
Beluga Whale	1000 – 123,000
Sea Lion	450 – 50,000
Harp Seal	950 – 65,000
Harbor Porpoise	550 – 105,000
Killer Whale	800 – 13,500
Bottlenose Dolphin	90 – 105,000
Porpoise	75 – 150,000
Dog	67 – 45,000
Cat	45 – 64,000
Rat	200 – 76,000

Table 1 – 2. Sound Frequency Ranges for Animals

The power threshold of hearing for humans is about 10^{-12} Watt/square meter. Table 1 – 3 shows various power levels of sound and hearing and shows that the range for humans is in the middle of sounds from quiet to very noisy.

Source	Power (W/m^2)	Intensity (dB)
Jet plane	10^2	140
Speech	10^{-6}	60
Rustling leaves	10^{-11}	10
Threshold	10^{-12}	0

Table 1 – 3. Sound Power Levels From Various Sources

The numbers in this table should be taken as illustrative only because sound levels vary at least with distance and frequency.

Funny Facts of Physics

The Very Large Scale

Let's consider this middle-like position of humans by starting with the very large scale. They live in a universe made up of many objects, especially galaxies, each of millions or billions of which may contain millions or billions of stars and planets. The Earth is in one galaxy called The Milky Way. In this galaxy the Earth revolves around one of its billions of stars. We call that star the Sun. The Milky Way is a medium size galaxy, the Sun is a medium size star, and the Earth is a medium size planet. The Earth is neither so close to the Sun that everything burns or boils nor so far away that all things on it freeze. With respect to our Sun, Earth is in what is called the Goldilocks Zone, a middle zone. You are consigned to seeing the universe from this middle position.

You can view some of the very large scale attributes of the universe by gazing at the stars at night. Telescopes gather electromagnetic radiations from other stars and galaxies. Humans have begun directly exploring the moon, the planets of our solar system, asteroids, and are preparing to visit mars. Explorations of the far reaches of the universe have been made and are continuing with unmanned probes. All other stars in our galaxy, however, will be out of our range of direct human contact and experience for a long time, and for other galaxies a much longer time if at all.

The Very Small Scale

Humans' experience is different for the very small scale of say atoms and subatomic particles. You can see stars, but you can't see or feel individual atoms. Moreover you can never directly explore and experience the world of an atom because you cannot shrink to atomic size. You can only directly experience atoms in their aggregated form when atoms and molecules combine to produce all the structures that you see in the heavens or on the Earth, including yourself. You can experience these small particles individually, in their non-aggregated form only indirectly through experiments with laboratory equipment.

In The Middle

Fundamental Dimensions
   Mass

The fundamental Dimension Mass involves mass, weight, momentum, and kinetic energy. The dimension mass and its companion weight are familiar ones which are measured in units like a few grams, thousands of pounds, hundreds of kilograms, and kilotons. But what do humans know of the mass of the Earth, about 6×10^{21} tons? The Sun has a mass of about 2×10^{25} tons, and black holes have masses from 10 times to 10^{10} times the mass of the Sun. Those are absolutely unfamiliar and inconceivable masses and beyond the range of direct experience of humans.

At the other end of the scale, atoms have masses of 10^{-31} kg. A mass so small humans cannot imagine it. Once again the range of their experience of the range of masses of the parts of the universe is only in the middle.

   Length

The fundamental Dimension Length devolves to the dimensions length, width, and height. Mathematically they are the three orthogonal (perpendicular to each other) coordinates usually labeled x, y, and z. Distances and sizes in everyday life are measured in units of miles, meters, feet, or centimeters.

On the very large end of the scale the size of the universe is given as billions of trillions of miles, or 30 billion light years. A light year is the distance light can travel in one year moving at the speed of 186,000 miles per second. Yes that is second not hour. Try grasping that size.

On the very small scale length dimensions are like 10^{-8} m for atoms, 10^{-12} m for nuclei, 10^{-15} m for protons and neutrons, and less than 10^{-18} m for quarks of which protons and neutrons are made. But scientists measure atomic and nuclear particle sizes in the unit Angstroms (ten millionth of a meter) or femtometer (a thousandth of a trillionth of a meter). It's almost impossible for humans to comprehend such small lengths. The lengths you are familiar with are in the middle of those from the very large to the very small in the universe.

Time

The fundamental Dimension Time involves time, velocity, and acceleration. The dimension time has two characteristics. One is named proper time. It is the time seen by an observer and measured by a clock at rest with respect to the observer. The other time is the time seen by the observer at rest of the time on a moving clock. The dimension time is typically measured in units of seconds, hours, days, and years. More about time is discussed in Parts Three and Five – Relativity and Cosmology.

On the scale of the lifetime of the universe times are of the order of 10^{20} sec. That is of the order of ten trillion years. It's difficult for humans to think and visualize even ten years ahead. How can you conceive of ten trillion years?

At the atomic and nuclear levels events happen very fast compared to the times of the lives of stars and galaxies. At the very small scale, times are of the order of 10^{-43} sec. Humans can think in terms of seconds, and they time races to hundredths of a second, but only with mechanical or electronic devices. It's impossible for them to think in terms of times less than tenths of a second, so how do they accept 10^{-43} sec.? The experience of time for humans deals with a scale in the middle of the very long and very short times in the universe. Again, you are in the middle of your experience of time.

Humans are inclined to say that time is universal, that time is the same all over the universe. Now is now, here and elsewhere. And humans can tell time by a clock, wherever the clock is, and whether they and the clock are standing still or moving. Time is time. Maybe not so. Funny things happen to time when dealing with very small or very large lengths, masses, or speeds such as those out of the range of experience of humans. Some of those funny things are presented in Parts Three and Five–Relativity and Cosmology.

In The Middle

Compound Dimensions

There are many dimensions compounded of the fundamental three dimensions. Some familiar ones are:

Speed

Speed (or velocity which indicates direction as well as magnitude) is a compound dimension made from length and time. Humans are accustomed to speeds of walking, running, automobiles, jet planes, and rockets soaring into space. The units in which they are measured are typically miles per hour, feet per second, or meters per second. Humans accept that geostationary satellites used for communication on Earth move at about 7000 miles per hour. But there are subatomic particles that move much faster. Then there is visible light, or any part of the electromagnetic spectrum, that moves at 186,000 miles per second. If we could bend light and nothing got in its way a beam of light could circle the Earth seven and a half times in one second. Can you visualize that speed?

Acceleration

This is the change with time of speed or velocity. If no force is applied to an object its acceleration is zero. Objects falling to the Earth because of gravitational force accelerate at g = 32.2 feet per second per second. The expansion of the universe is accelerating such that a galaxy one Mpc (Megaparsec) away recedes at 50 miles per second, a galaxy 2 Mpc away recedes at 100 miles per second, etc. (A parsec is about 3×10^{16} m so a Mpc is 3×10^{22} m). Your earthly experience of acceleration is in the middle of the very small and the very large possibilities.

Conclusion

You, like all human beings, are neither large nor small in size, mass, physical abilities, or range of your senses. You are somewhere in the middle, on a logarithm scale, relative to the rest of the universe from the smallest to the largest dimensions. Because of this position in the universe you are limited in your

Funny Facts of Physics

view of the universe, even with instruments, of the very large and very small scales. These limitations influence your understanding of the world and make some of your observations appear strange. These are the funny facts of physics, some of which are discussed here, and others are presented in the following chapters.

CHAPTER 2

WHY FISH DON'T FREEZE

In a room, warmer air rises toward the ceiling and colder air sinks toward the floor. Outdoors, air in ravines and other depressions is colder than the air at and above ground level. This phenomenon occurs because a volume of air will increase in size as its temperature increases and decrease in size as its temperature decreases. Because the mass of air does not change with temperature this means that its density–its mass per unit volume–decreases as its temperature increases.

Warmer air becomes "lighter" and rises above "heavier" air. Because a volume of air will decrease as it gets colder, cold air will sink because its density increases as its temperature decreases. More dense air sinks in the less dense air.

Almost all materials, including liquids, act that way. They shrink when their temperature is lowered. Shrinking means they keep the same mass but occupy less space or volume. That shrinking means their density increases. And more dense materials in a liquid will move deeper in the liquid. In that case as the temperature of a liquid drops to its freezing temperature it will start freezing at the bottom of the liquid, and freezing will continue to the top of the liquid.

This typical dependence of a liquid's volume on temperature is shown in Figure 2 – 1 for four different liquids. Figure 2 – 1 illustrates an almost linear relationship from about room temperature down to -273.15 C known as the absolute zero on the Kelvin temperature scale. Absolute temperature is discussed in Chapter 21.

Figure 2 – 2 shows the typical temperature dependence of the density of liquids on temperature.

Funny Facts of Physics

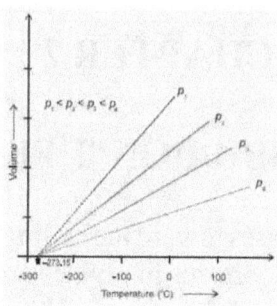

Figure 2–1. Liquid Volume versus Temperature

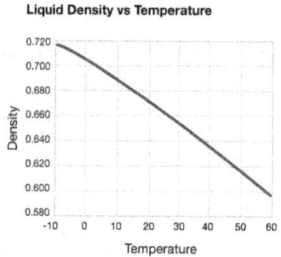

Figure 2 – 2. Liquid Density versus Temperature

A funny fact of physics is that not all liquids act this way, and that causes problems or offers benefits. Take water for example. You have noticed that in the winter when the temperature drops to and below 32 degrees on the Fahrenheit scale (32 F) or zero degrees on the Celsius (formerly known as Centigrade) scale (0 C), ice forms at the top of a body of water, say a lake. Why does it not form at the bottom of the lake as it does for other liquids?

As water cools it follows the pattern of other liquids and becomes more dense as it cools. As expected its more dense region sinks putting the colder water at the bottom and leaving warmer water at the surface. You may have experienced that

Why Fish Don't Freeze

when you dive down in a lake and sense that the water feels colder the deeper you go. And the water near the bottom of the lake is colder than near the top.

So, as water cools the colder parts of it sink. One would think that as it froze the ice would sink. The funny fact of the physics of water is that water becomes more dense as its temperature goes down until it reaches about 36 F or 4 C. That temperature is just about four degrees above the temperature at which water begins to freeze.

Here is where water shows its funny physics property. As it gets colder and drops below 4 C its density decreases rather than increases. That means it becomes less dense as it approaches and reaches its freezing temperature 0 C than it was just above the freezing temperature at 4 C. That means that as water cools below 4 C it begins to rise rather than sink. This peculiarity of water means that the coldest the water at the bottom of a lake can be is 4 C. That is until the entire lake freezes.

This behavior, unique to water, is shown in Figure 2 – 2 in which the volume of water is plotted against its temperature. It also shows the formation of ice. Figure 2 – 4 shows the temperature dependence of water density.

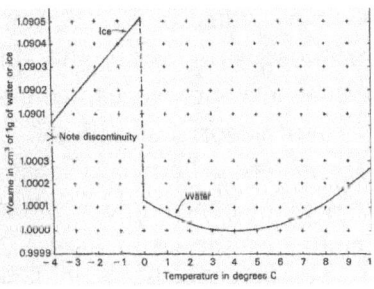

Figure 2 – 3. Water Volume versus Temperature

Funny Facts of Physics

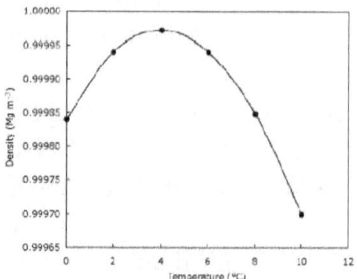

Figure 2 – 4. Water Density versus Temperature

That peculiar behavior of water is why ice forms at the top rather than at the bottom of a lake. That is good news for fish and anything that grows at the bottom of a lake. It is also a boon for ice skaters who don't have to skate at the bottom of a lake.

Another funny fact of physics is that Ambrose Bierce, an American satirist and journalist, wrote a definition of the word "ocean."

> "Ocean; noun; A body of water occupying almost two thirds of a world made for man–who has no gills."

Water has many other physical properties that differ significantly from those of other compounds with similar molecular structure. Those properties include melting point, boiling point, and heat of vaporization. Water has been said to be one of the strangest substances known to science. These features are attributable partly to the bond angle in water that differs from other similar molecules as shown in Figure 2 – 5. The bond angles in the other molecules are about 90 degrees, but in water it is about 105 degrees. That is about a sixteen percent difference between the bond angles that contributes to the peculiar properties of water. Now that is funny. Such a small difference between the arrangement of atoms in water and similar molecules accounts for water's very different behavior.

Why Fish Don't Freeze

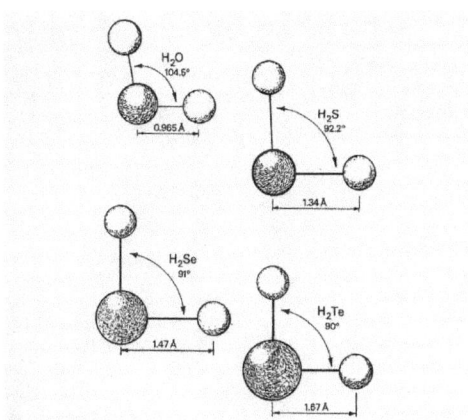

Figure 2 – 5. Bond Angle in Water

Conclusion

In a more serious vein though, the Earth would be a very different planet if water behaved like most other liquids because more than 70% of the earth is covered with water. If water did not decrease in density as it cooled below 4 C marine life would not have come into existence, oxygen would not be present in cold water, fish and crustaceans would not exist, and humans and their ancestors would have been deprived of an important food source. Humans would not exist. So you can attribute your presence in the universe to the funny, peculiar, characteristic of the change in density with temperature of water.

Funny Facts of Physics

CHAPTER 3

NOTHING IS BLACK

"I see that statue," you exclaim to a friend who is with you in the museum. But do you really see it? Think about that question.

When light, or any electromagnetic wave, from gamma rays to visible light, to radio waves shown in Figure 1 – 2, impinge on an object, some combination of four things happens to the electromagnetic waves. Here and in the rest of the book I will use the word "light" when it could be visible light or any other electromagnetic wave. The four possibilities of something happening to light when it strikes an object are:

1. The waves may pass through the object the way visible light goes through a pane of glass. That's called transmission.

2. The waves may be reflected producing a clear image like what you see when you look in a mirror. That's generally called specular reflection.

3. The waves may be diffusely reflected. A clear reflected image is not produced. It is blurry especially at the edges. Sometimes that is called scattering.

4. The waves may be absorbed the way infrared and ultraviolet waves from the sun are absorbed on your skin and give you a sunburn, or worse. That's called absorption.

Usually when light strikes a surface there is some probability of each of these four things happening. For example, most visible light is transmitted through a glass window pane, but you have noticed that under the right condition you also can see your image reflected in the glass. That is because the glass surface is relatively smooth. Some light is transmitted and some is reflected. Reflection from a rougher glass surface would be diffuse, not much transmitted, and the image would not be sharp.

Funny Facts of Physics

On a really sunny day you can feel the window pane warm up. That's caused by absorption of some light in the glass.

In many cases the lens surfaces of optical instruments and electromagnetic wave detectors are treated with a chemical coat to emphasize either reflection or absorption. Optical telescopes, for example, have antireflectiive coatings on their object lenses to reduce reflection and get more light transmitted to your eye, a photographic plate, or another type of light detector. Another example is the application on space helmets of reflective coatings on their face plates to reduce the amount of sunlight transmitted to the eyes of the space explorer. You have probably also seen sun glasses like that. Dark clothing is used to increase absorption of light.

Well what has all this to do with your seeing something and what is meant by black? The answer is that we see things by the light reflected or emitted from an object. Turn out the lights in the museum and you won't see the statue because there is no light to be reflected from the statue.

The funny fact of physics is that when you think you see a statue you do not see "it." What you see is the light that arrived at the statue from some source and that is reflected from the statue by varying amounts of red, orange, yellow, green, blue, indigo, and violet, toward your eyes. Light from the sun or from light bulbs is what is called white light. It is made of red, orange, yellow, green, blue, indigo, and violet light. The particular color of anything you see is because the object reflects that one color better than it reflects the other colors. As an illustration, if the statue appears white to you it is because the white light that impinged on it was reflected to you. You received all the colors. If the statue were to appear blue that would be because it reflected blue light better than the other colors.

If the statue was made of a material that did not reflect light you would see a black space with the shape of the statue as though the statue did not exist and there was a gap in the world

Nothing Is Black

with the shape of the statue. If you turned out all the lights in the museum you would not see anyone or anything in the museum because there would be no light to reflect from them to you.

If you were in a dark room and placed an opaque object near a wall, and illuminated the object with a light beam you would see a shadow on the wall in the shape of the object. Of course you recognize that the object intercepted the light so that the shadow is just a place where there is no light from the beam.

You see your shadow on the ground and conclude that your shadow is just a place with no light. A funny fact of physics is that shadows are not that simple. They are made of what are called an umbra and a penumbra which are shown in Figure 3 – 1 for a flashlight and an opaque object.

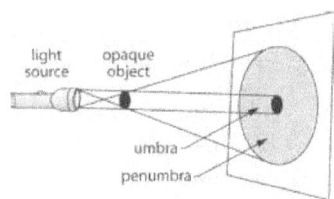

Figure 3 – 1. Umbra and Penumbra

Figure 3 – 2 shows the Earth, moon, sun system in which the umbra would represent a total eclipse and the penumbra a partial eclipse.

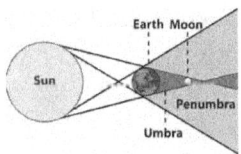

Figure 3 – 2. Earth, Moon, Sun With Eclipse

In addition to transmission, reflection, scattering, and absorption, another feature of all objects is that they emit infrared radiation called thermal radiation. The wavelengths and intensity

Funny Facts of Physics

of the radiation depend only on the objects' temperature. At room temperatures and human body temperatures this radiation is in the infrared portion of the electromagnetic spectrum. Human eyes cannot see it. Some animals and reptiles see infrared radiation.

However, if you wore infrared goggles, that capture and convert the infrared radiation into a visible color, you would see orange or green forms of the other people because their temperatures are at about 100 C. They would emit more infrared radiation than their surroundings which are at about 80 C.

And of course if an object blocks light, I. e., absorbs of reflects it back, a shadow forms behind the object which is just the absence of light. That is called–black.

There are two qualifications to all that I wrote above. The first is that of a television screen. The image you see on a television screen is formed from light emitted from the screen, possibly a cathode ray tube in an old TV set or a light-emitting diode (LED) or liquid crystal display (LCD) in newer models. That emission is somewhat like thermal radiation in that it comes from the object rather than from external light that is reflected or scattered from the object.

Conclusion

The word black really does not necessarily mean that nothing is there where you look. It means that no light that reaches it is reflected from it or scattered from it toward you. So the funny fact of physics is that black really can be thought of as nothing–no visible light reflected or emitted.

CHAPTER 4

ENERGY DOES NOT EXIST

People today talk a lot about energy. They and you worry about its availability and cost and the consequences of using it. Especially you are concerned about nonrenewable energy derived from the fossil fuels such as coal, oil, and gas. The desirable consequences of the use of energy are that energy provides people with warming when it is cold, cooling when it is hot, transporting things and people, and powering the myriad of mechanical and electronic devices people use.

The undesirable consequences of obtaining energy from nonrenewable fossil fuels are damage to the earth upon extracting it from the ground, the extinction of a natural substance, and the emission of pollution when the resource is consumed. Because of these and other problems humans today are in a quest to replace nonrenewable energy with renewable energy that is non-polluting.

Renewable Energy

Solar and wind energy installations have sprung up around the world, mostly in geographically remote areas because they occupy vast areas of land that are not available near metropolitan areas. These two renewable energy sources are receiving the most popular attention.

Nuclear fission energy that has been used for decades has increased in use around the world, but despite its recognized advantages it is not developing rapidly in the United States. The public still has an irrational fear of nuclear fission energy. This is despite the facts that no serious accidents have occurred and no one has been killed or badly injured in the United States from nuclear fission power reactors. A few governments and private companies have been working f or more than fifty years to produce the unlimited and non-polluting energy from nuclear fusion to generate electricity. The prospects for achieving it in the short term are not favorable.

What Is Energy ?

Rarely do people who talk so much about, argue about, and promote various types of energy ask themselves the simple question, "What is energy?"

Some people think it is a material substance that flows through one's body or objects, but energy is not a substance. No one can see, feel, smell, or weigh energy. The funny fact of physics about energy is that energy does not exist.

Energy is not a substance. It is a concept, an idea. I compare the concept of energy to that of the concept of credit. Credit is not a substance. Credit is merely a word that tells you how many real dollars are in a bank or how many things you can buy. Energy is a similar thought that enables you to numerically calculate the possible or actual movement of objects relative to each other or how temperatures may change. In other words, energy tells you how much work can be done.

The basic reason humans are so concerned about energy and have spent so much time and effort on the idea since the early cave men is that humans want something else–machines–to do their work. Now you have to ask, "What is work?"

The popular meaning of work is that it gets something done for you. Action takes place. Work, as defined in physics, however, is the numerical product of a force moving an object and the distance through which the object moves. For more details about the definitions of work and energy see Appendix 4.

Using this definition of work, a definition of energy that permits you to make numerical calculations of the use of energy, is:

Energy is a concept that allows you to calculate how much work can be done, or it lets you determine how much action can be taken.

Energy Does Not Exist

That's a reasonable definition because the Greek word for energy is ενεργεια which translates in English to "action."

If energy does not exist, what do the words nonrenewable and renewable energy mean? The funny fact of physics answer is that when you consider the use of energy you really mean the use of the resource from which you extract something that performs work. It is the resource, like coal, oil, gas, and uranium, that you consume. And these resources are in limited amounts on the earth so they are nonrenewable.

On the other hand wind and solar radiation are renewable sources of energy as long as the sun shines. Nuclear fusion energy is essentially a renewable source of energy because it involves hydrogen, which is in great abundance on the earth and throughout the universe.

Types of Energy

Energy can also be thought of as a property of a substance like its chemical composition, temperature, location, or velocity. So, though energy does not exist, it is convenient to think of it as a substance or an action. Then we get the familiar types of energy:

1. Nuclear – Fission, Fusion
2. Mechanical – Kinetic, Potential
3. Chemical
4. Electromagnetic Radiation, Electric Charges, Magnetic Poles
5. Wind
6. Solar
7. Heat

Conservation of Energy

A law of physics, known as the First Law of Thermodynamics, states that any type of energy can be converted to any other type and that energy is conserved. That means, for example, a calorie of chemical energy, say in a lump of coal, can be converted into a calorie of heat, no more and no less. A foot-pound of an airplane's potential energy (its altitude) may be

Funny Facts of Physics

converted into a foot-pound of kinetic energy (its speed) at a lower altitude. What is the energy that comes from a uranium atom bomb explosion? It is gamma and x- ray electromagnetic rays and kinetic energy of neutrons and atoms smaller than the uranium atom that are moving rapidly thus have lots of kinetic energy.

The Energy Ladder

All types of energy are equal, but a funny fact of physics is that some types are more equal than others. By that expression I mean they are more useful than others. The different types of energy reside on a scale of usefulness, like on a ladder, with the most useful type at the top and the least useful type at the bottom as shown in Figure 4–1. This fact is stated in physics as the Second Law of Thermodynamics.

Nuclear
Mechanical
Chemical
Electromagnetic
Wind
Solar
Heat

Figure 4 –1. The Energy Ladder

This Second Law of Thermodynamics states that it is easy to go down the energy ladder but hard to go up it. For example, rubbing your hands together is mechanical energy, and when you rub your hands together they get warm. That's heat energy. It is easy to go down the ladder from mechanical energy to heat. But heating your hands will not make them rub together. When you try to change heat to say mechanical energy–go up the ladder–some work must be done and some of the heat becomes not useful.

Energy Does Not Exist

This explains why coal, oil, and gas–nonrenewable sources of energy–were the first chosen sources of energy. It is easier to convert the chemical energy of coal, oil, or gas to heat. But to convert solar or wind energy–renewable sources of energy–to mechanical or electrical energy requires going up the ladder and takes work, some apparatus, and the loss of the use of some energy to things like friction and non-useful heat.

All this means that it is obvious why the use of renewable sources of energy may be less useful, less efficient, and possibly more expensive than the use of nonrenewable sources of energy.

Conclusion

In conclusion, though there is no quantity, energy, you can think of it as a material that flows back and forth and employ it as a substance to determine how much work you can get done. The caveat is that it is easier to go from a more useful type of energy like nuclear to a less useful type like heat–down the energy ladder– and more difficult to go the other way– up the energy ladder.

Funny Facts of Physics

CHAPTER 5

A NICKEL'S WORTH OF ENERGY

From the previous chapter you know that energy is not a substance. It is a property of a material, and it takes on many equivalent types from mechanical energy down the energy ladder to heat. And every type of energy can be converted to every other type. Now we show a funny fact of physics that mass and energy also are equivalent and can be converted from one to the other.

Energy–Mass Equivalence

Dr. Albert Einstein developed a theory of relativity that showed that time and space were different from what they seemed to be on an everyday scale. Some of the funny consequences of relativity I discuss in Chapters 10 – 14. One of Dr. Einstein's most famous predictions, and which most people have become familiar with and has almost become a household idea is his equation that relates and equates energy and mass, i.e, $E = mc^2$ where E is energy, m is the mass of an object, and c is the speed of light in a vacuum. His equation states that mass and energy are equivalent in that one may be transformed into the other, and that the amount of mass that is converted equals the amount of energy released divided by c^2.

A funny fact of physics is the unexpectedly huge amount of energy one can derive from a small quantity of mass. For example, think about a United States nickel, worth five pennies. It has a mass of five grams. The speed of light in a vacuum is 186,000 miles per second or 3×10^{10} centimeters/second. Putting these values in Eienstein's equation, results in
$$E = 45 \times 10^{20} \text{ Joules}.$$
A Joule is a unit of energy. In more familiar units E is equal to:

Funny Facts of Physics

    a. 1.25×10^8 kWh of electricity, or
    b. 15,000 tons of coal, or
    c. 100 kilotons of TNT.

To put item **a** in perspective, the mass in one nickel when converted to energy could supply electricity to 100,000 homes for a year.

For item **b,** one nickel could prevent you from burning 15,000 tons of coal to produce that electricity.

Item **c** represents the equivalent of the output of ten of the original uranium-based atom bombs. Recall that "energy is nothing" so what was the energy output of those atom bombs? It was in the forms of electromagnetic radiations like x rays and gamma rays and the kinetic energy of particles.

One other funny fact overlooked frequently is that Einstein's equation involves mass but does not specify any particular mass. The mass in Einstein's equation could be uranium, lead, air, beer, or anything even combinations of materials. It is the total mass, not what kind it is, that enters Einstein's equation.

Don't start hoarding nickels or lead or beer or other materials, though, expecting to get a great return on them in the future as sources of renewable energy. To date it is impossible to get the energy out of most materials without heroic activities consuming more energy than would be released. The only elements that presently are used to extract energy are the radioactive elements like the 235 isotope of uranium or the element plutonium. These elements are the sources of energy used in nuclear reactors.

Conclusion

According to Einstein's famous equation, all kinds of mass can be converted to energy, and with difficulty the reverse is true. The energy in a mass is astonishingly large and takes on many forms such as electromagnetic radiations and kinetic energy of released particles.

CHAPTER 6

YOU CANNOT GO IN REVERSE

You know very well that things cannot be done over. You cannot go back in time or reverse the flow of time. Everything moves in time in one direction we call forward, toward the future. When you perform actions or make changes in things those actions or changes cannot be reversed completely. They are essentially irreversible.

Our Actions are Irreversible

As an example, pour cream into a cup of coffee and with a spoon stir it clockwise. The cream and coffee mix. Now stir it counterclockwise. Does the cream separate from the coffee? That action cannot go in reverse. You cannot separate the cream from the coffee. Well that's not quite true. You can separate the cream and coffee possibly by running the mixture in a centrifuge. But that takes electricity which means you have to burn coal to supply the electricity. You cannot reverse the coal burning process, which produces ash and gases, and get the coal back.

A funny fact of physics is that in the world in which you live, you cannot go completely in reverse. Actions and events in the world you see and live in are irreversible.

You can think of many more illustrations of irreversible actions like the event in the Humpty Dumpty Nursery Rhyme, tearing a page in a book, getting sunburn, burning gasoline in your car, or posting something online. Once it's done, it cannot be completely undone. Light from a bulb streams out and cannot be reversed to flow back into the bulb. Radiations from TV transmitters flow out but cannot flow back into the transmitter. Electromagnetic waves cannot be reversed in time. They are irreversible.

Funny Facts of Physics

The Arrow of Time

Philosophers and scientists gave a name to this idea of irreversibility. They called it the Arrow of Time, and they subdivided that idea into four separate arrows of time.

Thermodynamic Arrow of Time.

Observations have shown that everything moves from a more ordered state to a less ordered state. An egg, which is in a highly ordered form can be scrambled, which is a disordered form. But a scrambled egg cannot be put back into its original ordered form.

Useful energy moves to a less useful state. It moves down the energy ladder shown in Chapter 4. For example rubbing your hands together produces heat, but adding heat to your hands does not make them rub together. Mechanical energy, like that of rubbing your hands together, is a more ordered form of energy than heat, which is the most disorganized form of energy. You run an electric current through a tungsten wire and it produces light, but shining that same kind of light on a tungsten wire does not produce an electric current in the wire. You burn gasoline in your vehicle and get exhaust products and heat that are less useful than was the energy in the gasoline. You cannot capture those exhaust products and heat and combine them to make gasoline. We are very familiar with but pay little attention to this thermodynamic arrow of time.

Entropy

This concept of order and disorder is encapsulated in what physicists call entropy. The more ordered something is the less entropy it has, and the less ordered it is the more entropy it possesses. On the energy ladder shown in Chapter 4, low entropy is at the top and high entropy is at the bottom of the ladder. This idea of entropy is so fundamental it is used in one statement of the Second Law of Thermodynamics. The use of energy with low entropy, which directly and without aid converts it (or related forms of energy) into a high entropy form of energy (or related

You Cannot Go In Reverse

forms of energy), is a thermodynamically irreversible process.
The Small Scale

A funny fact of physics is that on the small, the microscopic, scale the actions of small particles like molecules, atoms, electrons, and other nuclear particles can be reversible. The equations describing the motion of particles are the same for positive as well as negative time. This fact means the actions of the particles are reversible in time. Molecules in a collection obey these rules of mechanics. Therefor their motion is reversible in time. For example, a container of gas is made of molecules that are in constant random-like motion, and the system has a pressure and temperature. A mathematical description of the molecules would look the same if, in the equations describing their motions, time was positive or negative. In other words the molecules could reverse their actions completely. A movie of these molecules run in reverse would look the same as one run forward. If the molecules did in fact reverse their directions the system of molecules would look the same and have the same pressure and temperature.

The Large Scale

There is a conundrum here; a funny fact of physics. On the microscopic level actions are reversible, but though objects on the macroscopic scale and the cosmological scale are made of micro scale entities, the actions of objects on the macroscopic and cosmological scales are irreversible.

Psychological Arrow of Time.

You can remember the past but not the future. You cannot predict the future only speculate about it. You recognize that growing older, not younger, is a natural and irreversible process. This recognition of the thermodynamic arrow of time is manifested by the psychological arrow of time and is the one you are most attuned to. It dictates all you do and controls your life. By recognizing it you plan for the future and hope that your plans will materialize. You cannot undo things done, and you cannot reverse your aging process. Despite elixirs, pills, ointments, diets, exercise, and meditation that might extend the length of your life, you still can only grow older not younger. A longer life means you grow older for a longer time than someone with a shorter life. The psychological arrow of time is determined by the thermodynamic arrow of time. This is because you see things going from the ordered state to the disordered state (an increase in entropy) and all actions are irreversible. So everything you know about yourself and the macroscopic world is irreversible.

You might ask, "Is that how the entire universe works?" On the cosmological scale and on the everyday macroscopic scale the answer is, "Yes. Actions in our every day macroscopic and cosmological world are irreversible."

Cosmological Arrow of Time.

The best theory of the universe is that it started at a point in the vacuum with a Big Bang and has been expanding and evolving ever since. It is always moving toward the future. It cannot reverse in time and return to its origin.

There are speculations that the universe could collapse back in a Big Crunch. But that would not be a reverse of all that happened from the Big Bang to the time of the crunch. The collapse would still be an action toward the universe's future. Events would not reverse themselves. Cream would not separate from coffee. You, if alive then, would not grow younger. A Big Crunch would be as irreversible as was the Big Bang.

Quantum Arrow of Time

In quantum systems like a pair of electrons the transfer of heat has no meaning. What can be exchanged is information about the electrons, such as their spins. When such information is transferred the entropy increases. That process, like in the thermodynamic case, indicates a one way direction or an arrow of time.

Time Travel

An issue that has haunted humankind is whether or not time travel is possible. You might say, "The past happened. It exists. So shouldn't I be able to go back to the past?" On the other hand you say, "The future lies ahead. It will happen. So shouldn't I be able to travel to the future?" From all we know, some of which has been discussed in this chapter, you cannot go back in time, the arrow of time is always in flight in one direction.

But a funny fact of physics is that you can travel to the future. In fact you are always traveling to the future. At every moment of your existence you are traveling to the next moment, the future. As you read these words, you are traveling to the future. Actually you cannot stop traveling to the future. This fact is stated in the arrows of time.

Conclusion

The thermodynamic, psychological, cosmological, and quantum arrows of time fly only one way. You cannot go back in time. You cannot go in reverse. Though some events in the microscopic world are reversible, events in the everyday and macroscopic world are irreversible.

Funny Facts of Physics

PART TWO

GRAVITY

Funny Facts of Physics

How To Lose Weight

CHAPTER 7

HOW TO LOSE WEIGHT

You have heard people say, "Over the holidays I ate so much I gained ten pounds of weight." What they really mean is that they gained enough mass to weigh ten pounds more.

Mass and Weight

There is a difference between mass and weight. Mass is the stuff you are made of, the material that comprises your body and most other things. Weight, on the other hand, is a force. Weight is the force of gravitational attraction between your mass and the mass of the Earth. When someone says that they want to gain or lose weight they mean gain or lose mass so the force of attraction between them and the Earth is more or less.

In the English System of units weight has the unit pounds and mass has an unfamiliar unit called poundal[a] or worse yet, slug.[b] Weight and mass are related through the acceleration of gravity, g = 32.2 feet per second per second. The acceleration, g, is a description of the "g force" that flyers or astronauts are subject to in sharp or rapidly accelerating maneuvers. For an example of the difference between a person's weight and mass, a weight of 150 pounds results from the person's mass of

---

a. poundal is a unit equal to the force needed to accelerate a mass of one pound at a rate of one foot per second per second.

b. The slug is the unit of mass in the US common system of units, where the pound is the unit of force.

Funny Facts of Physics

150/32.2 = 4.66 slugs. Want to lose a female friend? Tell her, "You make a good-looking 4 slugs."

Weight on or Above the Earth

Weight, as explained above by Newton's theory of gravity, is the force of gravitational attraction between a person and the Earth. It varies directly with the mass of the person and the mass of the Earth and inversely with the square of the distance from the center of the Earth to the center of mass of a person's body. See Appendix 7.

From this theory, and the equation in Appendix 7, you can see that the farther you are from the center of the Earth the less you weigh by the square of the distance. So if you wish to lose weight the easy way is to just elevate yourself. You could walk up or use an elevator to go to a higher story in a building. Or you could climb a mountain. Or you could board a jet plane and soar to 30,000 feet. If you weighed 200 pounds at the surface of the Earth, you would weigh about 199.5 pounds at 30000 feet. And that weight loss would be temporary until you got back down. Using a jet plane to reduce your weight is hardly worth the effort and expense, especially because you would still have the same mass. You would be just the same size.

So the funny fact of physics is that using a jet plane or an elevator your weight would decrease but your mass, what you are made of, would not have changed. However, if you climbed the stairs in a building or hiked up a mountain to get farther from the center of the Earth that would decrease your weight a tiny amount, but because of the exercise you might lose some mass and subsequently lose weight. So gaining or losing mass is really what people desire if they wish to change their weight.

Weight Inside the Earth

Recall that r is the distance from the center of the Earth to your center of mass at the Earth's surface. What would happen to your weight if you went below the surface of the Earth in a mine, cave, or a science fiction machine that burrowed to the center of

How To Lose Weight

the Earth? You might think that because r is smaller you would gain weight.

A funny fact of physics is that when inside a mass the gravitational force on another object inside that mass depends only on the amount of the large mass from its center to that point inside where the other mass is located. The mass of the Earth M depends on its density and volume.

This means that the weight of a mass inside the Earth gets smaller as the mass moves deeper inside the Earth. The density of the Earth is not constant, but variations from it being constant do not change the qualitative fact that W varies as in the equation in Appendix 7.

So your weight, when inside the Earth decreases with r. The funny fact of physics is that your weight has its largest value at the surface of the Earth and decreases as you go away from or into the Earth. At the center of the earth you would weigh zero pounds. For a graph of your weight see Appendix 7.

The force of gravity varies inversely with the square of the distance between masses. Only at infinite distance does the force go to zero. What about the force when the distance goes to zero? The answer is that your weight would be zero in pounds or any other unit because there would be no mass of the Earth below you.

Conclusion

So you see the way to lose weight is to increase the distance between you and the center of the Earth above the Earth or to decrease the distance between you and the center of the Earth when inside the Earth.

Funny Facts of Physics

CHAPTER 8

THE MOON IS FALLING

A river rushes down a mountainside because, according to Newton's laws of gravity, the Earth tugs each drop, each molecule, of water toward the center of the Earth. The drops cannot go straight down because the mountain is in their way, so they slide along the mountain which is the only path available to them in their journey toward the center of the Earth.

Like the river, all objects, including you, fall toward the center of the Earth. Fortunately, in most cases, you are at the surface of the Earth and cannot fall any farther. Actually things are not pulled toward the Earth. According to Newton's Theory of Gravity the Earth and an object on, above, or in it experience a mutual force of attraction between the object and the Earth. You can equally say that you pull the Earth toward you as to say the Earth pulls you toward it. This mutual force of attraction with the Earth is called your weight. The attraction is mutual, but because the mass of the Earth is so much larger than objects on, near, or in it the Earth does not move as much as do the objects, including you.

Objects above the Earth fall toward the Earth. Throw a ball up and it falls back. Shoot a bullet horizontally and it travels some distance in that direction, but it is always falling toward the Earth. It travels only as far as it can in the horizontal direction until it falls to the ground. The speed with which an object held above the Earth falls is determined by the gravitational constant $g = 32.2$ ft/s/s, and the square root of the time it takes to fall some distance.

Using the equation in Appendix 8, if you drop an object from 6 feet it takes 0.61 seconds to hit the Earth. Starting at the

Funny Facts of Physics

same height, if you throw it horizontally, no matter how far it goes horizontally, it will hit the earth after the same 0.61 seconds.

The distance x an object would go in the horizontal direction depends on the speed with which you throw it and the time it travels. From the equation in Appendix 8, if you hold an object 6 ft above the ground and throw it horizontally at 10 mph = 14.7 ft/s the distance it will travel horizontally will be 8.97 ft.

Table 8 – 1 shows horizontal distances in feet that could be traveled by an object thrown horizontally at different speeds v and from different heights h. The left-hand column compares the numerical values of the speeds with common actions.

	mph	h = 6 ft	h = 100 ft	h = 1000 ft
Running	10	8.97	36.6	116
Auto	60	53.8	220	695
Jet Plane	600	538	2198	6951 ?
Sound	768	689	2814	8898 ?
Bullet	1700	1526	6228 ?	19695 ?
Rocket	6000	5384 ?	21981 ?	69511 ?

1 mph = 1.47 f/s

Table 8 – 1. Distance Moved vs Heights and Speeds

From Table 8 – 1 you see that if you project an object at six feet above ground horizontally at 6000 mph it will travel about one hundred feet more than one mile. For that distance the concept of horizontal loses its meaning because of the curvature of the Earth. That is why I placed question marks next to some numbers. When we say to propel something horizontally, we really mean at a tangent to a line from it to the center of the Earth as shown in Figure 8 – 1 for a circle representing the Earth.

The Moon Is Falling

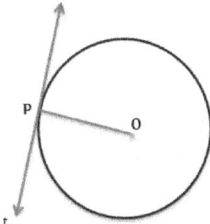

Figure 8 – 1. Tangent to a Radius of a Circle

Returning to Table 8 – 1 the items in the left column are not to be taken seriously. They are just to give a familiarity to the speeds in the next column. Obviously the table does not imply that jet plane at an altitude of 100 feet will crash after going 2198 ft. The plane does have lift from all of the other than vertical, surfaces.

Consider standing on a mountain 1000 feet high. If you drop an object, it will fall straight down because of the downward force of gravity. If you throw a ball in what I called above the horizontal direction, you are really throwing it at a tangent to the Earth's surface. When you do this action the ball will start along this tangent line but will also fall toward the Earth. Its resultant motion is an arc until it hits the Earth. It will always be falling at an angle to the tangent line. If you throw the ball harder, it will go farther, but eventually it will hit the Earth.

Now if you throw the ball a lot harder so it would travel at least say one mile before it hit the ground, something else influences its path. Thrown that hard the ball would still fall, but it would have to fall more than 1000 feet before it hit the ground. The reason is that the Earth is curved, and so as the ball moves, the Earth curves away from it. If you throw the ball even harder, the ball has to move even farther because of the continued curvature of the Earth.

Funny Facts of Physics

A funny fact of physics is that if you throw the ball fast enough so the distance it goes is equal to the circumference of the Earth (at 1000 feet high) the ball would keep falling but never hit the Earth.

That is what the moon is doing. It has enough tangential velocity to make it circle the Earth at its altitude above the Earth. That means though it is always falling, unless it slows down, it will never hit the Earth. See Figure 8 – 2.

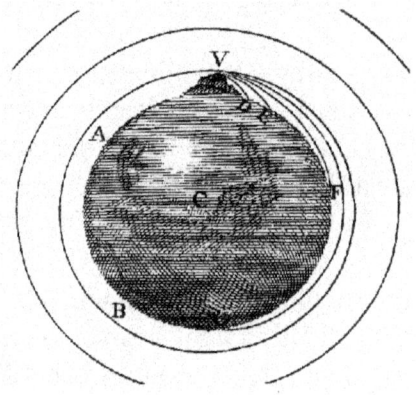

Figure 8-2. Trajectories of a Ball and the Moon

Figure 8 – 2 is taken from a work of Sir Isaac Newton and shows the various trajectories discussed above.

That the moon is falling toward the Earth is true not only for the moon but for any satellite circling the Earth. There is a potential problem for any satellite however. If a satellite is not high enough to escape all of the Earth's atmosphere, friction from the air will slow it down and it will fall to Earth or burn up in the atmosphere on its way down. Can that happen to the moon? No. There is no atmosphere at the altitude of the moon above the Earth to slow down the moon.

You might ask, "How fast do you have to propel an object

The Moon Is Falling

to have it go around the Earth, to be in orbit?" You know that if you tie a rock to a string and whirl it around the rock moves away and the string gets taut. That's from centrifugal force. You also know that all objects, are attracted to the Earth by the force of gravity. So if an object is going around the Earth and the centrifugal force is just equal to the gravitational force it should stay in orbit. See Appendix 8.

Now we see another funny fact of physics. The mass m is not in the equation for speed. It does not matter if you propel a BB, a keg of beer, an elephant, or a space station, the speed needed to stay in orbit at a distance ® + h) is the same for each of them. Getting each of those projectiles to the same speed, however, is another story. A lot more force and energy are needed to get an elephant moving at 100 mph than it takes to have a BB move at that speed.

Given in Table 8 – 2 are some speeds for an object to stay in orbit at different altitudes. From Table 8 – 2 you can see that if you are standing on the surface of the Earth, h = 0, you are moving in a circle at 17,718 mph. This speed and the others in the table may differ from values you see elsewhere because orbits are not circles so different calculations may use different values of h.

Orbit Type	r (miles)	h (miles)	r+h (miles)	Speed (mph)
Moon	3959	239000	243000	2,260
Geostationary	3959	22,200	26200	6,868
Low Earth Orbit	3959	99	4060	15,839
Earth Surface	3959	0	3960	17,718

Table 8 – 2. Speeds to Orbit the Earth at Different Heights

Funny Facts of Physics

Can the Earth act to slow down the moon's orbital speed? No, because a funny fact of physics is that just the opposite happens. The moon is slowing down the Earth's speed of rotation by 1.8 milliseconds per century. This loss of the Earth's kinetic energy is picked up by the moon which increases the distance of the moon from the Earth in its orbit. The moon actually moves away from the Earth 3.8 centimeters per year.

Conclusion

The moon is always falling toward the Earth, but it starts falling from a few centimeters farther away from the Earth each year.

CHAPTER 9

FOR A GOOD STRETCH TRY A BLACK HOLE

When you awaken from a long sleep or have been sitting a long time, like through a televised boring political debate, you sometimes feel you need a good stretch. Leap out of bed or the soft chair, get on your feet, and reach up and out and get that good stretch of your cramped and lethargic muscles.

If that's not enough you can find something to grasp and hang from. At worst you can journey to your favorite gym and do bends and hang from bars among other stretching pursuits.

Black Holes

Now let's digress. You've heard of and know something about what a black hole in space is. You also know about gravity and that all massive objects attract each other with what is called Newtonian gravitational force. As a reminder, a black hole is a very massive object in the universe, so massive that the gravitational force associated with it captures other things like dust, planets, and even stars into it. They fall toward it just like a thrown ball or the moon falls to the Earth except that after objects reach the "surface" of the black hole, called the horizon, they fall inside the black hole. A black hole is so massive that once objects go in they cannot get out. For a slight variation from that idea see Appendix 9.

Why They Are Black

Black holes have such a strong gravitational field that not only matter once in cannot get out but even light inside cannot get out. That's how black holes got their name. As explained in Chapter 3, you see an object only because light emanates from it or is reflected from it to your eyes. When that light enters your

Funny Facts of Physics

eyes you "see" the object. What you really see is the light coming from it. If light cannot escape from an object, the space it occupies looks black, hence the name black hole.

From the information given in Chapter 7, you know that the force of gravity on a mass outside another mass varies with the inverse square of the distance to that mass. The closer it is the stronger the force until it reaches the surface. Usually the difference of a few feet does not change the force very much. Now here is a funny fact of physics.

Stretched Like Spaghetti

A black hole has such a huge mass its gravitational force is especially large, and a few feet nearer or farther from it makes a large difference. For example, if you were standing just outside a black hole you would feel the force on your feet much more than the force on your head. The lower part of your body would be pulled toward the black hole with more force than would be the upper part of your body. The black hole would stretch you. And your body would compact from the sides. You would get long and thin. This phenomenon is whimsically called spaghettification and is illustrated in Figure 9 – 1.

Figure 9 – 1. Spaghettification

What would happen if you got pulled into the black hole? From Chapter 7 you know the gravitational force inside a mass is zero at the center and varies with the distance from the center. As

For A Good Stretch Try A Black Hole

you fell into a black hole more than likely the lower part of your body would feel less force than the upper part of your body but you would still get a good stretch. But a black hole is a different animal from a typical mass. Inside a black hole it would not only stretch you but even tear you apart, reduce you to your fundamental particles, molecules, atoms, electrons, protons, and neutrons. Now that's a really good stretch.

Inside a Black Hole

Despite that analysis we really do not know what happens inside a black hole. Our lack of an explanation is because both relativity and quantum effects may apply inside a black hole, but there is no relativistic-quantum mechanical theory. And no one has made experiments with a black hole.

Some speculations suggest that a black hole has no bottom. Things would fall through it and be tossed out into another part of the universe or even another universe. But another question arises. If that happened what would be the structure of what was emitted? If you fell in would you come out or would just the fundamental particles of which you are made come out? What we do know though is that getting near a black hole introduces huge effects of gravity unknown anywhere else in the universe.

Conclusion

If you want a good stretch a Black Hole can provide it, but you, like all matter and light, would not only stretch and be torn apart, but your parts would get swallowed by the black hole and never emerge.

Funny Facts of Physics

PART THREE

RELATIVITY

Funny Facts of Physics

CHAPTER 10

THE FASTEST

People love to see things moving fast and to have racing records set and then broken. A runner, boat, car, or plane can set a record and be called "the fastest." But in our world there is always the possibility of another runner, boat, car, or plane that at some time will go faster.

In every case, records for speeds take into account if there was a wind or water current moving with or against the moving person or object. Speed records are set relative to the ground so air or water motion with or against the moving object must either be added or subtracted to get the true ground speed. In the case of a sprinter setting a speed record, his or her time record may have time added because of a tail wind.

As another example when standing at the edge of a river, you've noticed a boat moving down stream, say at a speed of 10 miles per hour relative to the water. And the river is moving at a speed of 5 miles per hour relative to you. Then relative to you on the river bank the boat is moving at a speed of $5 + 10 = 15$ miles per hour. That's common sense and something you expect.

You have seen many other illustrations of this fact. Say a person in a train is walking forward at 2 miles per hour and the train is moving at 10 miles per hour. You, standing on the station platform would see the person in the train moving past you at $2 + 10 = 12$ miles per hour. Again that's just common sense and what you would expect. This idea of addition of velocities is accepted by you as normal. And there is no limit to the speeds of the object or the medium it is in. But there is a catch.

You have heard that according to Einstein's theory of relativity nothing can move faster than the speed of light given by the symbol, c. Now take a flashlight and turn it on. The beam of light moves relative to the flashlight at the speed of light, c = 186,000 miles per second, relative to the flashlight. That is so

Funny Facts of Physics

fast that if the light beam could bend and nothing got in its way it would go around the earth about seven and one half times in one second. Suppose while you are holding the flashlight you speeded up in the direction of the light beam to say ½ the speed of light. Like in the cases of the boat or train you would expect the light beam would be moving at a speed $1 + ½ = 3/2 = 1.5$ times the speed of light c, or 279,000 miles per second, relative to the ground.

Not so. A funny fact of physics is that the speed of light is always the same, c, relative to its source or any other observer regardless of the relative motions of the source or any observer. That is a consequence of Einstein's special theory of relativity.

Einstein derived an equation, like the one above, but for the addition of velocities near the speed of light called relativistic speeds or velocities. It is given in Appendix 10. If v is equal to the speed of light, c, then any value of u less than c will give a number for $w = c$. If $v = c$ then the number for w will be c. Even if we assume u is larger than c, w will always be equal to c. Even for $u = -c$, $w = c$. So in special relativity the funny fact of physics is that an observer can never see light moving faster than c even if the light is emitted by a moving source of the light traveling at a speed less than or equal to c.

Now take a case like we are accustomed to, say a plane moving at $u = 500$ miles per hour and a tail wind blowing at $v = 50$ miles per hour and $c = 186,000$ miles per second or 6.7×10^8 mph. What is w? As shown in Appendix 10, it is a number awfully close to 550 mph.

Conclusion

You see that in the real world, including that with all speeds up to that of light, nothing can move faster than light in a vacuum, and Einstein's velocity addition equation holds so that the speed of light does not change with respect to any observer. At speeds much less than the speed of light, the "common sense" equation explains our everyday world addition of velocities.

CHAPTER 11

FASTER THAN THE FASTEST

In the previous chapter we showed that nothing can move faster than the speed of light in a vacuum. That speed is usually given by the symbol, c. Even if the source that emits the light moves at any speed in the direction of the light beam or opposite to it the speed of the light beam emitted from the moving source as seen by an observer is always the same, c. This statement that there is a maximum speed, c, is a foundation of relativity theory that has been shown to describe accurately various features of the universe. It has been tested many times and has always proven to be true.

Faster than Light

A funny fact of physics, however, is that there are cases where something can exceed the speed of light. What is wrong? Is relativity wrong? Nothing is wrong if one accepts that the speed of light *in a vacuum* cannot be exceeded by objects with mass or even photons without mass. In other words no object or signal can move or send signals between points in the universe faster than the speed of light in a vacuum. The word light stands for an electromagnetic wave of any frequency and wavelength in the electromagnetic spectrum shown in Chapter 1.

Cerenkov Radiation

As an example of something that can exceed the speed of light consider the everyday scale of the universe. A funny fact of physics is that in a dielectric medium (an insulator that can have a positive electric charge at one end and a negative electric charge at the other end when in an electric field) something can travel faster than light in that medium. This phenomenon is seen by the emission of what was observed and explained by Vavilov

Funny Facts of Physics

Cerenkov. This Cerenkov radiation is electromagnetic radiation that is emitted when a charged particle, say an electron, moves through a dielectric medium faster than the velocity of light in that medium. So the electromagnetic radiation can move faster than the speed of light but not faster than the speed of light in a vacuum. See Figure 11–1 for an illustration of Cerenkov radiation.

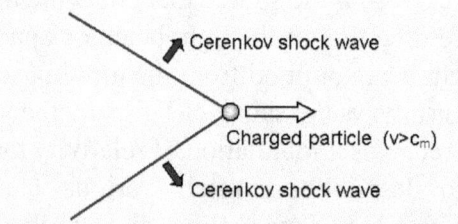

Figure 11 – 1. Cerenkov Radiation

In Figure 11 – 1 an electrically charged particle in a dielectric medium is shown moving to the right at a speed v which is greater than the speed of light in the medium c_m. The electromagnetic wave emitted is much like the shock wave from a jet plane moving faster than the speed of sound in the medium (air).

Expanding Space

Another example of the statement that something can move faster than the speed of light involves the large scale of the universe. Galaxies are known to be moving away from each other, and it is concluded that the universe is expanding, and its rate of expansion is increasing. If this phenomenon continues eventually the speed at which galaxies move relative to each other could conceivably exceed the speed of light. Galaxies are made of massive objects like stars and planets. How could large galaxies with their huge masses exceed the speed of light? What is wrong?

The funny fact of physics answer is that it has been

Faster Than The Fastest

proposed that galaxies are not moving away from each other but that space itself is expanding and taking the galaxies along with it. It's sort of like spots on a balloon seeming to move away from each other as the balloon expands. Because no mass or photons are involved in the expansion of space and no signals are being sent, space could expand at a speed greater than the speed of light in a vacuum. This phenomenon would not violate the fundamental concept in relativity that no mass or electromagnetic signal can exceed the speed of light in a vacuum.

Future astronomers and cosmologists will be unhappy about this phenomenon because eventually the galaxies could be so far from each other and moving so fast light from one galaxy might not reach another galaxy. Astronomers would have fewer objects to look at and study, and they would not be able to measure the continued expansion of the universe. On Earth, to you the sky at night would look almost empty because you would see only those objects in our milky way galaxy.

Conclusion

There are cases where something can move faster than light. But experiments and observations have shown that Einstein's postulate that neither matter nor electromagnetic radiation can move faster than the speed of light in a vacuum holds true. This postulate is a fundamental part of relativity theory.

Funny Facts of Physics

CHAPTER 12

SHRINK HAPPENS

You're coasting along at the speed limit in your car, and a car passes you on the left at a constant speed obviously exceeding the posted speed limit. You stare at the car and its driver who stares back at you. Nothing strange happens to either car from your or the other driver's point of view. Well not really.

If the driver of the passing car had the car moving near the speed of light things would change. A funny fact of physics is that as the car rushes past you moving at a constant relativistic speed (near the speed of light) something strange happens. At this speed the person in the moving car feels nothing different, but the moving car and its driver appear shortened to you. See Figure 12 – 1. On the other hand, your car looks shorter to the other driver though to you your car has not changed.

Figure 12 – 1. Relativistic Length Contraction of a Car

Noticeable changes in the length of a moving object will take place, for an outside observer, if the object moves at a constant relativistic speed. In special relativity physics this is called length contraction. It is as they say, relative. You think the other car is moving at high speed. The other driver could just as easily feel that your car was moving and his car was standing still.

Funny Facts of Physics

You may have experienced this feeling when sitting in a train and another one pass the one you are on. You can't tell which one is standing still and which one is moving. At lower speeds, like those to which we are accustomed in our everyday world, length contraction happens but is so small it is negligible. For instance in the case of you in the train, the train passing at nonrelativistic speed does not shorten noticeably. But at a relativistic speed it would appear to you as in Figure 12 – 2.

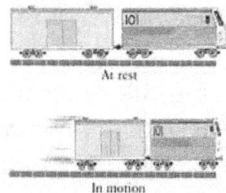

Figure 12 – 2. Relativistic Length Contraction of a Train

At the ballpark you follow a pitched ball from the mound to the plate traveling near 100 mph. With good eyes you can watch its path. And the ball never changes.

But if the pitcher had enough juice to get the ball moving at relativistic speed and it accelerated toward the plate, from the side you would see something different. The ball would start looking to you like a circle. As the ball speeded up it would look like an ellipse. At even faster speed you would finally think it looked like a vertical line. This contraction of a baseball would look like that in Figure 12 – 3.

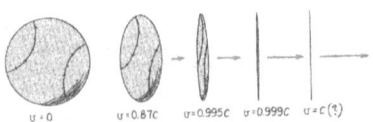

Figure 12 – 3. Relativistic Contraction of a Baseball

Instead of a car or train or baseball take a simple thing like a board. To measure the length of a board call the end points x_1

Shrink Happens

and x_2. If the person doing the measurement and the board are stationary the length L is just x_2 minus x_1.

Einstein and others showed that if the board starts moving and reaches relativistic speeds the observer who is stationary will see the length of the board getting smaller. To the observer the board length will look like L_o is L/γ where the Greek letter γ is always larger than one. His equation describing length contraction is shown in Appendix 12.

Values of γ and the fraction of length contraction of an object an observer sees, versus various speeds of the object, are shown in Figure 12 – 4 and Table 12 – 1.

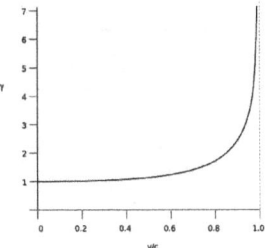

Figure 12 – 4. Graph of γ vs. Speed Relative to c

Then the equation becomes $L_o = L/\gamma$ or $\gamma = L/L_o$ which is the relative shortened length of the board seen by the stationary observer. The quantity γ is always greater than one, so the stationary observer sees the board as less than its stationary length L.

Values of the quantity γ and the fraction of length contraction of an object an observer sees, versus various speeds of the object, are shown in Table 12 – 1.

For example, from Table 12 – 1 a car passing you at 0.99 times the speed of light (184,140 miles per second) would appear to you to be 14.1 % of its stationary length.

Funny Facts of Physics

Speed of car in units of the speed of light c	γ	L_o/L
0.05	1.001	0.999
0.1	1.005	0.995
0.5	1.155	0.866
0.7	1.400	0.714
0.9	2.294	0.436
0.99	7.089	0.141
0.999	22.37	0.0447
0.9999	70.71	0.0141
0.99999	223.6	0.00447

Table 12 – 1. Relativistic Length Contraction

Conclusion

There is an apparently strange, from our everyday common sense point of view, feature of Einstein's theory of special relativity. In general it is that the length of an object moving at a constant relativistic speed appears, to a stationary observer, to contract along its direction of motion. Someone at rest (relative to the moving object) would see the moving object become shorter. It may seem funny but an observer on the moving object would see the length of the other observer grow shorter also. After all, motion is relative.

CHAPTER 13

HOW TO STAY YOUNG

Idling in your car at a red traffic signal when you are in a hurry you pound the wheel, "Won't this light ever turn green?" Time seems to drag and run awfully slow. A youngster in school stares out the window at the sunshine and fluffy clouds. It's the last day of school before summer recess, and he glances again at the clock praying for it to hit 3:00 when school will end. "Why is it running so slow?"

Then there are times when having fun and you glance at a clock. "It's that late already. Time sure went fast."

Everyone knows people feel time runs slow when they are eager for it to pass or fast when they want it to slow, but they all know that time does not run fast or slow. Or does it?

In previous chapters I discussed the dimension time and its features and peculiarities and some funny facts about time. In all that I wrote, however, I never suggested that the passage of time was anything more than a constant, that it runs neither fast nor slow. Time "flows" the same from the past to the present to the future, and we can measure its passage. And time is the same for everyone at every moment, everywhere, and under all conditions. Or is it?

People thought the same about space and the length of objects. Space was constant, it could not change, and the length of any object in it was the same no matter who looked at it and how fast it moved. Chapter 12 presents a discussion of how Dr. Albert Einstein showed us that thinking of the length of moving spacial objects as constant was wrong. The funny fact was that the observed length of a moving object depends on how fast it is

Funny Facts of Physics

moving and the motion of the person who views it.

Einstein went further and showed another funny fact of physics. He showed that time also was not fixed, and that things like clocks ticked along at different rates depending on their motion relative to observers of the clocks. This is called time dilation.

Let's say a man is standing still and a woman is moving at relativistic speed v relative to the man. The man sees a time interval Δt_s on his watch. The time interval Δt he sees on the woman's watch is longer than his time interval. In other words, from his observation, her clock seems to run slower than his. This dilation of time of a moving object is $\Delta t = \gamma \Delta t_s$ or $\Delta t_s = \Delta t / \gamma$ as shown in Appendix 13.

Values of γ and the change in the time interval of a watch seen by a stationary observer, versus various speeds of the watch, are shown in Figure 12 – 4 and Table 13 – 1. Figure 12 – 4 shows that, because γ is always larger than 1, the stationary observer sees the stationary time interval as longer than the moving time interval.

Speed of Woman's Watch in Units of c	Time Interval of Stationary Watch (Min)	Time Interval of Moving Watch Seen by Stationary Man (Min)
0.05	15	14.98
0.1	15	14.90
0.9	15	6.50
0.99	15	2.12
0.999	15	0.67
0.9999	15	0.21
0.99999	15	0.07

Table 13-1. Time Dilation

How To Stay Young

Values of time dilation for different speeds of the moving watch are shown in Table 1 – 1.

Twin Paradox

This time dilation effect led to what was called the Twin Paradox that puzzled many persons. It goes as follows. Suppose there are identical twins, a man and a woman each 40 years old. The man stays on the earth and the woman leaves in a rocket ship near the speed of light, say 0.99c. The rocket ship journeys out for say ten years and then returns taking ten more years. Because of the phenomenon of time dilation, when the traveling twin steps out of the rocket ship the man aged 20 years so is 60 years old, but from the table his traveling twin sister aged only 14.1% of those twenty years and is only 42.8 years old.

Real Examples

A few examples of the use of, need for, and experiments with time dilation are given below.

A. Clocks on GPS satellites

Satellites used to locate position on the earth would not be as accurate as they are without corrections for time dilation effects.

B. Muon decay

Muons are sub-atomic particles created when cosmic rays strike the upper atmosphere. They have a half life of about 2.2 microseconds (µs) meaning that every 2.2 µs their population is cut in half. By counting the number of muons at the top and the bottom of a mountain, we can see what proportion of them have decayed according to time dilation.

C. Caesium clocks

Planes with caesium clocks aboard flew in east and west directions along the equator circling the earth at different altitudes. Clocks on the eastward trip lost time and clocks on the westbound trip gained time compared to stationary laboratory clocks.

Conclusion

Funny Facts of Physics

Time is not absolute, and, like the case for lengths, it varies according to the relative speed between sources and observers of time measuring devices like clocks or living things. A stationary observer will see a moving clock running slower than the stationary clock or the moving person aging slower than the stationary person. So if you want to stay young, travel. But you have to do it near the speed of light which now and in at least the near future is impossible.

CHAPTER 14

HOW TO BULK UP

You want bigger muscles, more bulk, more mass? To get bigger muscles you don't have to spend hours in a gym pressing massive weights and stuffing yourself with vitamin and health drinks or, worse yet, steroids. All you need to do is walk faster. Well maybe. It is true, but you would have to walk really fast, say close to the speed of light to show any appreciable gain in mass.

A funny fact of physics is that the mass of an object increases with its motion. The faster an object moves the greater its mass becomes. The process is similar to that for length contraction of objects and time dilation discussed in Chapters 12 and 13. The equation for increase of mass with speed is given in Appendix 14.

To illustrate the change in mass with speed Table 14 – 1 shows the ratio of the mass of an object relative to its rest mass at speed v. Note from Figure 12 – 4 and Table 14 – 1 that v has to be about one half the speed of light, that would be 93,000 miles per second, before an appreciable change in mass occurs. So don't count on using speed to help bulk up your body now or in the near and even distant future.

Particle accelerators used for atomic and nuclear particle experiments can make atomic particles like electrons reach relativistic speeds, and their mass increases according to Einstein's equation as shown in Table 14 – 1.

For example a proton moving at 0.9 times the speed of light would have a mass 2.294 – almost two and a half – times its mass when it is at rest.

Funny Facts of Physics

v in units of c	$\gamma = m/m_0$
0.05	1.001
0.1	1.005
0.5	1.155
0.9	2.294
0.99	7.089
0.999	22.37
0.9999	70.71
0.99999	223.6

Table 14-1. Ratio of a Moving Mass to its Rest Mass

From Chapters 10 and 11 you learned that nothing can move faster than the speed of light in a vacuum. Objects with any mass, no matter how small, can never even reach the speed of light. One of the reasons it is impossible for an object with mass to reach the speed c is because, according to Einstein's theory of special relativity, as you approach the speed c the mass increases toward an infinite mass. That means that an infinite force would be needed to move an infinite mass. That, of course, is impossible.

Conclusion
So you can bulk up by moving but you have to get going at relativistic speeds to see any appreciable increase. Nothing we know of can move you or other macroscopic objects that fast, so if you really need bulk, use the old-fashioned methods of exercise and nutritious food.

PART FOUR

QUANTUM THEORY

Funny Facts of Physics

CHAPTER 15

PARTICLE OR WAVE

Photoelectric Effect
Dr. Albert Einstein received a Nobel Prize for deciphering the photoelectric effect. This effect is the emission of electrons from a metal when light is shined on the metal. For illustration, consider a case when low frequency light, such as the color red, is shined on a metal. No electrons are emitted no matter how high the intensity of the light. However, when green light is used, electrons emerge with a certain velocity (kinetic energy). When blue light is shined on the metal, electrons emerge but with even greater velocity. The greater the intensity of either green or blue light the more electrons will be emitted.

Considering light as a wave no one could explain the effect. Dr. Einstein succeeded in explaining the phenomenon by assuming light was not a wave but was made of packets of energy $E = hf$ where h is Planck's Constant and f is the frequency of light when it is thought of as a wave. These quanta of energy were called photons and can be thought of as massless particles.

In this way Einstein showed a funny fact of physics that light could be seen either as a wave advancing in a medium or as a stream of particles. The choice depended on how one observed the effects of light.

Pinhole Camera
You know how a pin hole camera works. Light passes through a small circular hole in a box and impinges on a photographic film where it creates an image. Because not all the light goes through the exact center of the hole, the image will be blurred at its edges. So the smaller the hole the sharper the image.

Funny Facts of Physics

The hole need not be circular. It could be a slit, but if so the image produced would in the shape of the slit.

Single and Double Slit Experiments With Light

Replace the camera with a flat screen with a thin vertical slit in it, and put a photographic film behind the screen. Let a coherent light beam[c] pass through the slit. Most of the light will pass through the center of the open slit making a big spot on the film. Less light will pass through off the center line of the slit and be recorded at the edges of the center of the image like for a pinhole camera. The image on the photographic film would look like the image in Figure 15 – 1 which represents the intensity of the light. You would expect this pattern if light was a wave or if it was a stream of photons.

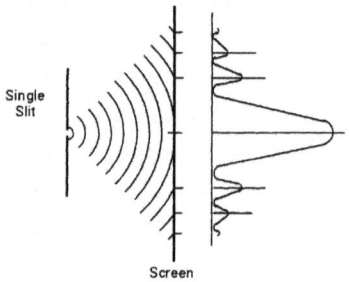

Figure 15 – 1. Single Slit Pattern for Light

Now put a second slit in the screen parallel and near the first slit. Close the first slit and pass the light beam through the new slit. The pattern on the film will look like that in Figure 15 – 1 except it would be behind the second slit. Now open both slits and let the light beam pass through. You might expect two

---

c. Coherent means that all waves in the beam have the same frequency and are in phase with each other.

Particle Or Wave

patterns on the film each alike but each behind each slit. A funny fact of physics is that this is not what you would see. The pattern would be like that in Figure 15 – 2, a series of bright and dark regions with no central maximum.

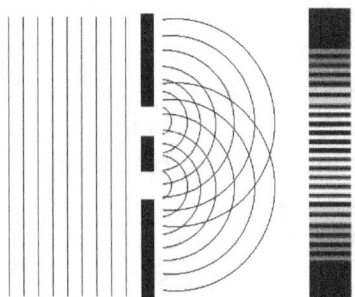

Figure 15 – 2. Double Slit Pattern for Light

This behavior can be explained by considering light as made of waves that spread out after they pass through each slit and then interfere with each other constructively and destructively causing the bright and dark lines. You would not expect this pattern from the massless particles called photons. Einstein's explanation of the photoelectric effect experiment showed that light was made of particles called photons. This two slit experiment, however, demonstrated that light can be considered as a wave.

Single and Double Slit Experiments With Bullets

If bullets, which are macroscopic objects, are fired through a single slit they form a pattern like that for light. If both slits are open, they form two patterns like the single slit pattern with each pattern behind one of the slits. That's what you would expect and that's what you see.

Single and Double Slit Experiments With Electrons

Complementing Einstein's idea that light can be a stream

Funny Facts of Physics

of particle-like photons, Louis de Broglie suggested that particles like electrons might act as waves. So we have two ideas that waves can act like particles and particles can act like waves. These suggestions had to be resolved.

Now do the same experiment that was done with light except instead of using bullets use small particles, say electrons, and an electron detector replacing the film. Shoot a beam of electrons at the screen with one slit open. The image on the detector will look like that in Figure 15 – 1for the one slit experiment with light. Close that slit and open the other slit and shoot the electrons in. The pattern at the detector will be the same as for the other slit except at the location behind the open slit just like for light. All this seems reasonable and expected.

Now open both slits and shoot in the beam of electrons. Because electrons are considered particles, common sense dictates that the individual electrons will pass through either one slit or the other. You might expect to get the pattern from bullets fired through two slits. Then there would be two patterns like that in Figure 15 – 1, one behind each slit on the detector. A funny fact of physics is that for electrons the pattern will look like that in Figure 15 – 2. This implies that the electrons acted like waves and interfered, like light waves do, after passing through the slits.

If you think this two slit experiment with electrons is funny consider propelling electrons toward the screen with the two slits open one electron at a time. Surely you would expect each electron would pass through one or the other of the slits so they should contribute to the single slit pattern for the single slit they went through. Another electron or photon might go through the other slit and then contribute to the single slit pattern for that slit. That makes common sense.

But. Eventually, after many electrons are used the bright and dark interference pattern shown in Figure 15 – 2 will appear. The same is true for light if you let light through the two slits one photon at a time. But, you think, this is strange. A single electron or single

Particle Or Wave

photon can go only through either one slit or the other, and if they go through at different times they can't interfere with each other.

But a funny fact of physics is that even when photons or electrons pass through the double slit arrangement one at a time each photon or electron acts as though it goes through both slits at the same time and the experiment produces the two slit pattern. Now that seems funny, but it's true.

The funny fact of physics is that light, or any electromagnetic radiation, and electrons, or other small particles, may be considered to be either waves or particles depending on how they are experimented with or how they are observed. For a time physicists could not explain this apparent conundrum and some of them wanted to call waves or particles wavicles. After further development of quantum theory that nomenclature was deemed unnecessary.

Experiments, other than the two slit experiment, have confirmed that under some conditions light can be thought of as particles called photons. Under other conditions light acts as waves. Similarly other experiments have shown that small particles can act like particles or waves depending on the experiment.

Like most physical phenomena, funny, or strange, or not, there is an explanation. This time the explanation is in the realm of quantum theory. In this theory a particle or photon is not thought of as a thing like a billiard ball or as wave in space or in a medium but as a wave function (The concept of a wave function is discussed in Chapter 16). That's funny but consider this. The wave function does not represent the particle or wave, but the square of the amplitude of the wave function represents the probability that the electron or photon will be at a particular place at a particular time. The wave functions for photons or electrons in the two slit experiment are not localized but spread so that the wave function of each photon or electron can be thought of as passing through both slits.

Funny Facts of Physics

Conclusion

Electromagnetic waves and atomic or subatomic particles can exhibit the properties of waves or particles depending on how you observe or experiment with them. Their properties and actions obey the rules of quantum mechanic wave functions that give probabilities of their properties.

CHAPTER 16

SCHRÖDINGER'S PATCHWORK

Equations in physics, including wave equations, can be derived from physical laws. For example, the wave equation for mechanical vibrations on strings, drum heads, air columns, and in many other cases can be derived from the laws of mechanics. The functions that are solutions of the equations are expressed mathematically as waves that represent the displacement of matter when forces are exerted on it. As another example, electromagnetic waves are solutions to Maxwell's equations in which the wave solutions can be electric and magnetic fields varying in space and time.

During the development of quantum mechanics people searched for an equation that would sum up some of the experimental evidence and other partial theories of the actions of small particles. The experiments and theories developed by Einstein and others about the duality between particles and waves implied that a suitable equation would be like known wave equations. For a long time no one found such an equation.

A funny fact of physics is that a suitable equation with waves as solutions was discovered, but it had no relationship to classical wave equations. Dr. Erwin Schrödinger patched one together from ideas about conservation of energy, and he found solutions to it that were wavelike.

The basis for Schrödinger's equation is the energy of a particle or collection of particles, and it is a separate postulate of quantum mechanics. The solution of his equation is a wave function that describes the particle or the whole system. The

Funny Facts of Physics

Schrödinger equation was a new and bold idea. Richard Feynman, commented:

> "Where did we get that (equation) from? Nowhere. It is not possible to derive it from anything you know. It came out of the mind of Schrödinger."

What caused Schrödinger to generate this equation? From the explanation of the photoelectric Einstein concluded that light could be made of photons. Louis de Broglie suggested that the reciprocal situation might exist and particles might act like waves. The two slit experiment provided good evidence that both ideas were correct. These ideas furthered the development of quantum mechanics and led to speculations about how to mathematically, by equations, describe these quantum phenomena. Among the many attempts was the successful one by Erwin Schrödinger. He followed the Newtonian idea that a particle in motion had kinetic energy K and potential energy V that added up to a total energy E or

$$K + V = E.$$

He formed an equation that looked like that simple one, but he postulated that a particle had to have wave properties so the solution to his quantum equation had to be wavelike. There are many equations in physics that have wave solutions, all of which can be derived from the physical facts and observations. In other words they describe physical models.

A funny fact of physics is that Schrödinger saw that quantum physics of particles had no model, and particles were not waving around but they were sort of in a wave field. The equation he formed had no relationship to known wave equations. Essentially he patched one together that included the coordinates x, y, and z of a particle, its mass m, time t, and the potential energy of the particle V and gave a wave form as its solution. It is explained in Appendix 18.

Now another funny fact of physics is the inclusion of the

Schrödinger's Patchwork

imaginary number $I = \sqrt{-1}$ in Schrödinger's equation. How can a real particle and its change in space and time have an imaginary aspect? The answer is that the appearance of I produces a mathematical solution that has wave characteristics. For example
$$e^{I\pi} = \cos \pi + I \sin \pi$$
which is sinusoidal like a wave.

Another funny fact of physics is that the wave function does not represent any properties of the particle, but the square of the amplitude of the wave function represents the probability that the particle will be at a certain place at a certain time.

There is a related time independent equation Schrödinger constructed by replacing the last term by the energy of the particle multiplied by the wave function. It is shown in appendix 18. The solutions to this equation give the discrete (quantum) energy levels the particle can have. They are called eigenvalues.

These two equations started a new vision of quantum mechanics and was found to work extremely well for many physical problems. So the funny fact of physics is that a postulated equation, which could not be derived from known physics, introduced a new idea of particles being not just waves but being some kind of wave function. The nature of those functions is still being debated, but the equations work for solving particle and other problems.

Conclusion

Schrödinger introduced an equation not derivable from physical laws but nonetheless introduced the idea that a particle in motion can be expressed mathematically by a wave function. The wave function does not represent the position of a particle in space and time, but the square of its solutions can be used to give the probability of a particle being somewhere at some time.

Funny Facts of Physics

CHAPTER 17

YOU CAN NEVER BE CERTAIN

Everyday Events

In our everyday life we live in a macroscopic (large scale) world in which it is possible to be certain about all kinds of facts. For instance we can be certain at any moment to simultaneously know where something is and how fast it is moving.

For example, an air traffic controller scans the screen. The screen shows the coordinates of one jet 10.2 miles away at 110 degrees, altitude 5,080 feet, and speed 412 knots on a course of 145 degrees. This may be one of a dozen planes being tracked, and their coordinates, altitude, speed, and direction must be known exactly and simultaneously. Imagine the problems if the controller said, "The jet is 10.2 miles away at 110 degrees and 5080 feet, but I'm not sure of its speed." Speed and position must be known exactly at the same time. And they can be known to a certainty whose dimensions are at most those of another jet to avoid collisions.

As another example, Jeanette, using her smart phone, calls her friend Charlie in his automobile driving to her house. Jeanette asks, "Where are you and when will you be here?" In other words she is asking him for his position and his driving speed at a particular moment.

Charlie answers, "I'm at exit 32 and going 65 miles per hour. Should be there in an hour."

One can determine that a moving automobile can be instantaneously at a specific place and simultaneously determined to be moving at a specific speed. There is no uncertainty about those two simultaneous measurements. This principle of "certainty" is common sense for us because it holds for all

Funny Facts of Physics

relatively large, macroscopic, objects and at speeds familiar to humans.

Small Scale Events

For much smaller objects (on the microscopic scale), however, a funny fact of physics is that one cannot make these simultaneous precise measurements of certain properties of a particle such as position x and momentum p, which is velocity (speed) multiplied by the mass m of the object. The more accurately one measures position the less accurately can one determine the momentum of a moving object, and vice versa. Simultaneous answers to these measurements can be made but only to some uncertain degree in the measurements. This result is not caused by uncertainties in the instruments or measuring procedures. It is a fundamental funny fact of physics.

This phenomenon is expressed in what is called Heisenberg's Uncertainty Principle. It states that simultaneous measurements of position and momentum of an object have an inherent uncertainty. The two quantities are inversely related. That is, the more accurately one measures one of the properties the less accurately one can measure the other property. The equation for Dr. Werner Heisenberg's Uncertainty Principle is given in Appendix 17.

Large Scale Events

If this small world situation applied to the large world Charlie, in his automobile might answer Jeanette's request with, "I'm going between 60 and 70 miles per hour and might get there between six and eight o'clock, and I'm at exit 32." Or he might say, "I'm going exactly 65 miles per hour and will get there at exactly seven o'clock, but the best I can tell you is that I am somewhere between my house and yours."

Let's see how this works for another everyday situation. Consider a baseball thrown from a pitcher's mound 90 feet from the plate with an uncertainty of 4 feet. When it strikes the catcher's mitt, it is measured as traveling 100 miles per hour.

You Can Never Be Certain

Using the uncertainty principle equation you get $\Delta v = 6.3 \times 10^{-34}$ mph..

That means the uncertainty you have about the ball moving at 100 miles per hour is only an incomprehensibly small fraction of 100 mph. So even though the location of the baseball in the catcher's glove has an uncertainty of 4 feet the speed of the ball measured as 100 miles per hour is for all practical reasons exact. For details see Appendix 17.

Small Scale

Now consider the really small scale world of a proton with a classical radius of 10^{-15} m and mass = 1.67×10^{-27} kg moving at one half the speed of light. We can measure its position with a detector. Say it has an uncertainty, Δx, in this measurement of a magnitude equal to its radius, 10^{-15} m. A simultaneous measurement of its momentum will have an uncertainty of $\Delta v \geq 3.16 \times 10^7$ m/s

This mean that if one measured the position of a speeding proton to within a distance of its radius, the uncertainty in its velocity would be about 10^7 m/s or about 20% of its speed. For details see Appendix 17.

Energy – Time Uncertainty

Another funny fact of physics is that just as there is an uncertainty between the simultaneous measurement of the two properties position and momentum of a particle, there is a somewhat similar uncertainty between other properties. One of them is between the property energy of a particle and the external variable time as shown in Appendix 17. This form of the Uncertainty Principle has a number of interpretations. As one example in this equation time t during an experiment is fixed or increases and has no uncertainty in its value. The quantity Δt in this case is the time it takes for a system to change its energy significantly which is given by the uncertainty in the change in energy of the system, ΔE.

Funny Facts of Physics

Conclusions

All measurements on physical systems, including simultaneous measurement of related variables, have some inaccuracy or uncertainty because of things like instrument errors. In addition, though, there is a more fundamental uncertainty in the values of some simultaneous measurements such as those of position and momentum of an object or energy and time of a system. These inherent uncertainties are negligible for macroscopic objects but can be large for small scale things like protons.

CHAPTER 18

SOME ACTIONS ARE SPOOKY

In our everyday world all objects, even a collection of similar objects, that are not intentionally connected mechanically or electrically are independent of each other. The properties of one, such as its state of motion, are independent of the similar properties of the other objects. And measurements of the properties or actions of one object do not affect the values measured of the other objects. This fact is more pronounced the greater the distance the objects are from each other.

For example, hold two baseballs in one hand and throw them at the same time. Somehow measure the way one is spinning and then measure the other one. The way one spins does not influence how the other one spins. Once the balls leave your hand they are independent of each other and are subject to air currents and other factors. You might say the baseballs retain their independent properties at that point in space and at that time where and when each is located. Physicists call this aspect of the world locality.

A funny fact of physics is that independence of the properties or states of similar objects starting together is not necessarily true for small particles, like photons, electrons, or protons, that obey the laws of quantum theory. Physical systems like two electrons or two photons started together can be intertwined with each other. Physicists call this condition quantum entanglement. The state of one electron or photon can be determined by the state of the other one because of this entanglement. If you measure some property, like spin, of one electron the value of the measurement of spin of the other

Funny Facts of Physics

electron is fixed by the measurement made on the first one.

The same applies to say the polarization of photons. This is true even over large distances, as much as many kilometers. Even if not two but many small particles are entangled, measuring the quantum state of one particle constrains the value of the measurements of the quantum states of the other particles. Physicists call this phenomenon nonlocality. In 2017, using a Chinese satellite named Miscius, a physicist named Jian-Wei Pan beamed entangled photons to two cities 1200 km from each other.

Dr. Albert Einstein found this phenomenon of entanglement hard to believe, and he called it "spooky action at a distance." If a measurement on one object influenced the other object, it was proposed that a signal had to have been sent instantaneously from one object to the other. That action meant the signal would go faster than light. And that was shown by Einstein to be impossible. Some of this was discussed in Chapter 10.

This idea presented the possibility that entanglement and nonlocality violated the fundamental idea that nothing can move faster than light in a vacuum. Einstein and his colleagues Dr. Boris Podolsky and Dr. Nathan Rosen, constructed Gedanken (thought) experiments to disprove this spooky action. It is referred to as the Einstein, Podolsky, Rosen, or EPR, paradox.

Unfortunately for the three men, Niels Bohr demonstrated that they were wrong. Despite doubting it by at least those three eminent physicists, quantum entanglement and nonlocality were proven to be aspects of nature. This idea of quantum entanglement, or nonlocality, has since then been confirmed in many experiments that justified the theoretical proof of its existence. The entangled objects do not have local properties. Their properties are non-local.

The explanation for nonlocality is that particles are not little objects or traditional waves but are wave functions that satisfy a fundamental equation given by Erwin Schrödinger discussed in Chapters 15 and 16. These wave functions can spread over distances involving many similar particles each of which can be in a different state.

One explanation for entanglement is that upon making a measurement on one particle the wave function involving all particles collapses into a single state that determines the properties of all the particles no matter how far they are from each other (i.e., the particles exhibit nonlocality).

Conclusion

With our limited view of the world and how it works, this action of entanglement and nonlocality of the properties of electrons and other very small particles does seem spooky, but in the quantum world of the small it happens.

Funny Facts of Physics

CHAPTER 19

BATHED IN PHOTONS

Water and Air Oceans
Fish live in an ocean of water. Humans live in an ocean of air. Fish, and even humans, can swim up and down and float in the water ocean, but neither fish nor humans can swim or float in the ocean of air. You can float in water, but you sink in air.

Why Things Float
 The reason you sink in the air ocean and can float in the water ocean is because the mass density of air is about one thousandth the mass density of water. For something to float in a fluid–a liquid or gas–it must displace an amount of fluid whose weight is equal to or greater than the weight of the object. If the object's weight is concentrated in a small volume (the object has a large density) the volume of fluid displaced will not weigh as much as the object, and the object will sink. But if the weight of the object is spread over a large enough volume (the object has a small density) the volume of fluid displaced will weigh more, and if the weight of the fluid displaced is equal to or greater than the weight of the object the object will float. You've seen pictures of weather balloons filled with helium rising and floating in the air. That's because helium is lighter than air, and the balloon plus helium weigh very little compared to their large volume so they displace a volume of air that weighs more than the balloon plus its helium. Huge cruise ships weigh about 220,000 tons but they displace about 500,000 tons of water. That's what keeps them afloat.

Electromagnetic Ocean
 In addition to the oceans of water and air, you live in an

ocean of electromagnetic energy. You are bathed in this electromagnetic energy ocean. The electromagnetic energy spectrum ranges from gamma rays at the high energy end through x rays, ultraviolet, visible light, infrared, microwaves, and radio waves at the low energy end. See Figure 1 – 2 for a picture of the electromagnetic spectrum and Appendix 19 for more on electromagnetic radiation.

This ocean of energy is made of waves, or we can consider it made of photons, which have essentially no mass thus no weight. So the mass density of any object, including you, is vastly larger than that of the electromagnetic energy ocean so that swimming or floating in the energy ocean is impossible. You sink.

You are surrounded by ultraviolet, visible, and infrared light from the sun and light bulbs and infrared radiations from our surroundings such as walls and furniture and living things. Radio and television broadcasts produce radio waves and microwaves. Satellites and our personal electronic gadgets, phones, tablets, computers, GPS devices, radars, and navigation beams for ships and aircraft exchange microwave radiations. And the electromagnetic waves are always there. Turn on your cell phone almost anywhere and you pick up and use or add to those radiations.

You are bathed in a spectrum of electromagnetic radiations. Each of these types of electromagnetic radiation has a different frequency and wavelength related to the speed of light (which means the speed of all electromagnetic waves). High frequency means short wavelength and low frequency means long wavelengths. You are bathed in this electromagnetic energy, battered by it constantly.

There is an interesting feature of the components of this electromagnetic radiation spectrum. It is that instead of thinking of the components as waves scientists have proved that we may equally regard them as made of streams of particles called

Bathed In Photons

photons. This realization is discussed in Chapter 15. Each photon has an energy appropriate to its frequency when it is considered to be a wave. See Appendix 19. So a funny fact of physics is that you are bathed in and are constantly pummeled by a swarm of tiny bundles of energy called photons coming from all directions. Some photons, like gamma ray photons, have enough energy to pass through your body, x-ray photons may deposit in internal human tissue, ultraviolet photons damage your skin, visible light photons allow you to see objects as mentioned in Chapter 3, infrared photons produce warmth, and radio and microwave photons are used for electronic communication.

Ionizing Radiation

Of the various forms of electromagnetic radiation are those that can damage materials including your body. As mentioned above, they are the higher frequency gamma rays, x rays, and ultraviolet radiation. They cause damage because they have enough energy to knock electrons out of the atoms from which materials and human bodies are made. Physicists refer to this action as ionization.

Nonionizing Radiation

The other lower frequency radiations from visible light through infrared, radio, and microwave radiation usually do not have sufficient energy to knock electrons from atoms. They are non ionizing radiations and are less harmful.

Conclusion

Though photons have energy and momentum, you should notice that the quantity mass does not appear in any of the equations about them. Photons have no mass. It is fortunate for life on earth, especially humans, that photons have no mass. If photons did have mass, you would be bombarded constantly by a variety of different energy particles, like tiny billiard balls, knocking you about. That scenario might not be too comfortable.

Funny Facts of Physics

CHAPTER 20

EXCITING LIFE OF A PHOTON

Among a swarm of trillions of photons that streamed out into space at 186,000 miles per second and started their 2.5 million-year journey to you there is a handful of special ones. All of these special photons started this journey just so you can point to a spot in the night sky and say, "There's the Andromeda Galaxy. It's the nearest spiral galaxy to the Milky Way and contains a trillion stars." A picture of the Andromeda Galaxy is shown in Figure 20 – 1.

Figure 20 – 1. The Andromeda Galaxy

Of those trillions of photons, spread across the energy spectrum from gamma rays to radio waves, with many in the visible region, not all of them will make it to you. At least three groups, however, say in the middle part of the electromagnetic spectrum you identify as the color red, will survive the trip.

For a photon it is a difficult and dangerous road that stretches about ten million trillion miles from the Andromeda Galaxy to Earth. The route is cluttered with all kinds of space objects from other large galaxies to tiny protons, which could

Funny Facts of Physics

gobble it up or deflect it from its path. Clouds of interstellar atoms, molecules, and dust block the path. The solar wind, which consists of charged particles released from the Sun, lays in wait. The interplanetary magnetic fields and the Earth's magnetic field are awesome but present no direct barrier to the speeding photons. But those magnetic fields influence the concentrations and paths of charged particles that could interact with the photons. Stars, planets, nebulae, clouds of hydrogen atoms, planetoids, asteroids, dark matter, comets, black holes, and the whole spectrum of electromagnetic radiations rumble around in space. Over millions of years of their journey, free atoms absorb some photons which disappear as excitations of the atoms.

Now, to all those obstacles, add the thousands of operating human launched satellites and the half a million pieces of human caused debris that orbit the Earth.

For thousands and tens of thousands of years other photons from Andromeda are scattered by free electrons and electrons bound in atoms, so those scattered photons no longer sail toward you. During the next million years while the ones headed toward you fly past massive objects like galaxies or black holes some are deflected from their paths and sail off in other directions.

A funny fact of physics is that trillions of trillions of photons actually do fight their way through this jungle of small particles, huge masses, magnetic fields, radiations, and human obstacles and reach Earth. Among the surviving photons are three particular groups of photons. After a million years, maybe only a few trillions of the original stream of photons are left on their original path. The three groups of interest continue toward the Earth. Dust, debris, and small and large space objects steal many more photons from the original swarm. But, after passing through these trillions of miles and journeying for millions of years and having most of their companions devoured, scattered, or deflected, the three special bunches are still streaking earthward.

Exciting Life Of A Photon

The Earth's atmosphere halts the paths of many more photons, especially the blue ones that are scattered. Among those that finally arrive at the Earth are our three special collections of photons.

Now near the end of their arduous trip they strike the objective lens of your telescope. Your eye is almost glued to the eye piece at the other end of the telescope. The smooth outer surface of the objective lens tosses some photons back out into space. They are reflected. The atoms of the glass snag a few percent of the photons when the glass absorbs them. The third bunch of photons zip through the objective and eye piece lenses of your telescope. They are transmitted and end their life absorbed in the retina of your eye. "I see Andromeda," you exclaim.

Among one of the three special groups of photons that was reflected from the outer lens of the telescope one of them flashes upward to 30,000 feet where it bounces downward off the wing of a speeding jet plane. A minute fraction of a second later it scatters from a pitched baseball into the stands and reflects sideways from a fan's eyeglasses out toward the city. There it scatters from a dust shrouded window on the 40th floor of a skyscraper. In a flash it burrows through the open window of a first grade school room. There it reflects off a kid's brightly colored metal lunch box and exits a different window heading heavenward. Next it bounces off a port hole of the space station and heads for outer space. Fifteen minutes later it impinges on Venus where it is scattered into deep space.

Assume that just by coincidence this photon's trajectory is on a perilous path, dodging dust, ions, atoms, and galaxies, to where Andromeda will be somewhat less than 2.5 million years from now and some thousand trillion miles closer to and on a collision course with the Milky Way. There in Andromeda, after surviving the dangers of space for five million years and its encounters with the Earth, the photon will arrive just as fresh and potent as when it started its journey. Most probably, then, it will

Funny Facts of Physics

end its romantic and exciting life by being gobbled up by and giving its energy back to an atom in one of the Andromeda Galaxy's four hundred billions of stars or planets.

But if it does not end its romantic five million year journey by returning its energy to the electronic structure of an atom in a star or planet in Andromeda, it surely could tell an exciting story.

Conclusion

You see an object in the night sky like a galaxy and think, "There it is." You have to remember, though, that the light from it started millions or billions of years ago and fought its way through large and small items that make up the universe to you. What's more is that every single photon contributed to your sighting of that galaxy, and every single photon had a history of its own traveling around the universe.

CHAPTER 21

COLDER THAN THE COLDEST

The two familiar temperature scales you use in everyday life are Celsius (formerly named Centigrade) and Fahrenheit. The scales and the number and sizes of the degrees in each scale are based on two properties of water, its freezing and boiling points.

On the Celsius scale the freezing temperature of water is defined at one atmosphere of pressure as 0, and the boiling temperature of water at one atmosphere of pressure is defined as 100. So there are 100 Celsius degrees between the freezing and boiling temperatures of water.

On the Fahrenheit scale the freezing temperature of water is defined at one atmosphere of pressure as 32, and the boiling temperature of water at one atmosphere of pressure is defined as 212. So there are 180 Fahrenheit degrees between freezing and boiling temperatures of water.

To go from one temperature scale to the other means shifting them by 32 to get the same starting point then adjusting the size of the degrees by the ratio of 180 to 100. For example 70 F is the same temperature as 21 C. And 15 C is the same temperature as 59 F. See Appendix 21.

Notice that for temperatures on the F scale below zero one gets negative F degrees of temperature. They are called "below zero Fahrenheit" temperatures. Similarly for temperatures on the C scale below zero one gets negative temperatures. They are called "below zero Celsius" temperatures.

Absolute Temperature

A question arises. How far below zero can either the C or

Funny Facts of Physics

F scale go? How negative can they get? Or how cold can something get?

The numbers 0, 32, 100, and 212 seem strange and you may wonder, "Why choose them? Why not some other physical properties of water or some other substance and then give other numbers to them?" Most likely the reason is that water is abundant, easy to work with, and its changes of phase from ice to water to gas are within our everyday temperature experiences. But the F and C scales give awkward numbers, and again the question arises, "How low can a temperature be?" To answer that question above, you first have to know what is meant by the concept of temperature.

Physicists define temperature as the speed of motion of atoms or molecules in a material. The faster the particles move or the higher their speed, say in miles per hour, the higher the temperature of the material. Their top speed is unlimited in principle, but a molecule's least speed is in principal zero miles per hour. If all the molecules in the material had zero velocity that would represent the absolutely lowest temperature of the material. Some say that is where all motion ceases and there is no energy in the material. Quantum mechanics makes a correction to that idea.

With that in mind physicists defined two other temperature scales called Absolute Temperatures using the F and C scales that each start at zero. The absolute F scale is called Rankine, R, and the absolute C scale is called Kelvin, K. In the R scale at one atmosphere of pressure water freezes at 491.69 R and boils at 671.69 R. In the Kelvin scale at one atmosphere of pressure water freezes at 273.15 K and boils at 373.15 K. The temperature scales are shown in Figure 21 – 1.

Colder Than The Coldest

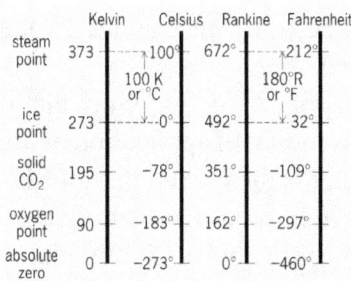

Figure 21 – 1. Comparison of Temperature Scales

On these more meaningful R and K absolute scales there are no negative temperatures. Absolute means just that. By the definition of temperature and in the absolute temperature scales, absolutely no material can have a temperature below zero. On the basis of the definition of temperature a negative temperature would mean atoms or molecules have negative speeds. That is meaningless, therefor, negative temperatures cannot exist.

Negative Absolute Temperature

That is true in our everyday world but what about on the microscopic quantum level? A funny fact of physics is that on the quantum level it is possible to have what is called a negative absolute temperature. In a collection, by that I mean a population, of atoms some are in the ground state energy level and others are in excited energy states or levels. Usually there are more in the ground energy level than in the first higher energy level and fewer in the subsequent higher energy levels. At all times some are moving up and some are falling back between energy levels, but in equilibrium and when undisturbed the normal numbers in each energy level stay the same. For a diagram and discussion see Appendix 21.

Now suppose the atoms can be radiated enough so that the number of them in the higher level becomes larger than the

Funny Facts of Physics

number of them in the lower level.. In that case for the equation in the appendix to be true, the absolute temperature T would be considered negative.

Can this happen? Yes. A funny fact of physics that has seen a tremendous amount of application of this idea of a negative absolute temperature is the invention of the laser. The word laser is an acronym for light amplification of stimulated radiation. A laser operates on the principal of illuminating say a crystal of ruby so that the atomic energy levels of chromium in the ruby crystal are inverted. There are more Cr atoms in the high energy state than in the low energy state. This is not an equilibrium situation and if the atoms are stimulated to move down to the lower energy level they will simultaneously emit a coherent light beam, a laser beam.

Conclusion

You see applications of this idea of a negative Absolute Temperature every day in science, medicine, military, entertainment, and commercial uses. Lasers, which operate on an inversion of populations of energy states of atoms, are the best well known applications of this idea of a negative Absolute Temperature. It is a temperature below the lowest temperature or it is colder than the coldest.

So a funny fact of physics is that in the microscopic quantum world one can have a situation that is described by a negative absolute temperature. That is a temperature colder than the coldest which is zero.

CHAPTER 22

THE VIRTUAL VACUUM

The womb of the universe was a vacuum. How can something, especially something as immense as the universe, be born from a vacuum, from nothing? A funny fact of physics is that a vacuum is not nothing.

Space

For a long time most people thought outer space was a vacuum. In the last century astronomers and others have discovered that the region between interstellar objects, such as galaxies, normally called space, is really a very crowded place. Between the galaxies there are clouds of hydrogen, helium, neutrinos, cosmic rays, other particles, and electromagnetic radiations and fields. Space also is bathed in what is called the cosmic background radiation (CBR) which has a frequency spectrum of that of a Black Body at a temperature of 2.7 K. For a discussion of a Black Body and the CBR see Appendix 22. This CBR is called the after glow of the Big Bang. It is the electromagnetic radiation left over from what was emitted shortly after the Big Bang at a very high Black Body temperature and has changed its Black Body spectrum as the universe cooled.

The Vacuum

Space is far from being a vacuum, but a funny fact of physics is that even if all the objects and radiations were removed what would be left still would not be a vacuum. So you might ask, "What is a vacuum? Why did I title this chapter a virtual vacuum? How did the universe come into being from a vacuum, a virtual vacuum?"

Some theories used for answering those questions suggest that the vacuum can be an entity that is populated by what are

Funny Facts of Physics

called virtual particles. They are pairs of particles and their antiparticles that appear for such a short time they virtually do not exist. Virtual particles appear courtesy of the Heisenberg Uncertainty Principle in the form $\Delta E \, \Delta t \geq h/4\pi$ as given in Chapter 17. This equation asserts that particles may come into existence for short times Δt because of uncertainty in their energy ΔE. The virtual particles absorb this energy for a very short time then give it back to the vacuum. And this process repeats and repeats. Because the particles have such short lives, they are not permanent thus they are called virtual particles.

Big Bang Theories

Other theories suggest these particles are matter and anti-matter pairs which, when they recombine, release tremendous amounts of energy as radiation and particle motion. The theory postulates that the violent recombination of matter and anti-matter resulted in the Big Bang.

Another way of looking at the occurrence of the Big Bang and the appearance of all the matter in the universe from nothing is that it satisfies the First Law of Thermodynamics which states that energy is conserved. And a funny fact of physics is that the total energy of the universe is zero. How can that be?

The answer is that matter can have positive and negative energy. The energy of motion of particles is positive, but the energy of gravitational and electromagnetic fields is negative. These positive and negative energies balance out to zero. The universe started from nothing and is still, in sum, nothing.

Conclusion

The vacuum from which the universe formed was not nothing. It was filled with virtual particles whose fleeting, virtual, existence might have been responsible for the creation of the universe. The universe, vast as it is, may be thought of as a "zero sum" item in that its positive and negative energies sum to zero.

PART FIVE

COSMOLOGY

Funny Facts of Physics

CHAPTER 23

I SEE A STAR

On a dark clear night you gaze up and see a sky full of lights you call stars. Some of them are stars like our sun which is an ordinary star. Some of those points of light you view may be stars that are our neighbors in our own galaxy we call the Milky Way. Some of them may be planets that, like the Earth, are circling the sun and reflect the sun's light to you. Other points of light may be galaxies similar to our Milky Way galaxy, each containing millions of stars, that are so far away they seem like points of light. You think of them as stars. So let's call all of the objects you see stars because that's what they look like.

How far away can some of these objects, these stars, be? To answer that question, first let's set some distance standards and units. Light travels at 186,000 miles a second or 3×10^8 meters per second. The distance light travels in one second is defined as a light second. The distance light travels in an hour is a light hour. And so on. Because distances in the universe are so large, cosmologists, astronomers, and physicists use as a distance standard the distance light travels in a year. They call this distance a light year. It is the distance light would travel in one year at 186,000 miles every second of that year. Because there are different definitions for the length of a year, there are slightly different values for a light year. Typically the equivalent of a light year is about 5.88×10^{12} miles or 5.88 trillion miles.

Cosmologists, astronomers, and physicists have studied these heavenly objects with optical, radio, infrared, x-ray, gamma-ray, and other kinds of telescopes. A conclusion they drew, from all these different electromagnetic radiations they see and how those radiations change randomly or periodically, is that they can

Funny Facts of Physics

see from Earth out to a distance of about 13.7 billion light years. That distance amounts to 8×10^{22} miles. That is about ten billion trillion miles.

Now, of all the millions of stars you see, you pick one star out which happens to be a galaxy containing maybe a million stars 10 billion light years away from Earth and say, "I see that star." Really? The funny fact of physics is that you don't really see that star. What you see is light from that star that started toward you 10 billion years ago, about six billion years before Earth formed. That means that the star you think you see may now be somewhere else than where you are looking, and it may be different from what you see, or it may not exist.

How could it be somewhere else than right where you see it? Well during the lifetime of the universe more galaxies were born and other matter gathered together creating changes in the gravitational attractions among galaxies so they might shift positions. Because of gravity the objects in the galaxy, you see as a star, may coalesce and have fewer but larger stars. Then too, the universe has been expanding since it started as a Big Bang, so the star, that galaxy, was closer to you when it was younger than it might be now.

Why could it not exist? It's possible that during that 10 billion years after the light you see now started its journey to your eye, that galaxy may have merged with another one, or the whole galaxy may have collapsed into a Black Hole. You can find out if any of those events happened if you are patient. Wait another 5 billion years and that new light that started toward you 5 billion years ago will then be in your view. At that moment you will see the star where it was and what it was, not ten billion years ago, but only 5 billion years ago. That sight will tell you what happened to it during at least that 5 billion-year interval.

Conclusion

When you look up at the sky at night, remember that the

I See A Star

stars and the whole part of the universe that you see may not be what you think they are. What you are looking at is light that started from those objects maybe millions or billions of years ago. Now, those objects may be at some other location, have changed their structure, or maybe not even exist.

Funny Facts of Physics

CHAPTER 24

TIME WILL TELL

You have grown accustomed to the idea of time. You live your daily life according to time schedules for rising, eating, playing, going to school or work, and slipping into bed for a night of sleep. You plan events days and months in advance. Travel by bus, train, ship, and plane is according to strictly prearranged time schedules. You accept and measure time's passing with clocks. Often you think about time that has gone by and what could have been. You speculate on future time and what it might bring. Clocks and calendars are ubiquitous symbols of your involvement in, fascination with, and reliance on time. Time is an integral, essential, and unavoidable part of your life, everyone's lives, and of the entire universe.

    Archeologists have uncovered many human remains and human-made artifacts that they declare were constructed millions of years in the past. Other explorers have discovered fossils of animals that lived six hundred million years ago. Astronomers have dated the ages of the earth, the sun, and other planetary bodies as billions of years. Cosmologists have gone even further telling us that the universe we see with our eyes, optical telescopes, and other detectors of radio waves, gamma rays, and x rays is close to 14 billion years old. Scientists of all categories have made calculations and observations of these various ages and periods of the Earth and its inhabitants and are in general agreement with the conclusions. So is there a problem?

Earth Years

    The problem is that these dates and time intervals are given in units of years. By that I mean Earth years. What is an Earth year or simply a year? It is the time it takes for the Earth to

Funny Facts of Physics

make a complete circle around the sun. That's one orbit of the sun by the Earth. It is also the time it takes for the sun to be in a particular position, say its lowest point called a winter solstice on December 21-21 then for the Earth to go through the spring equinox on March 21-22, the summer solstice on June 21-22, and the fall equinox on September 22-23 until the sun appears again in that same winter solstice position as shown in Figure 24 – 1. Then you can dissect that year into twelve parts called months or about 365 parts called days or about 8760 parts called hours and so on down to seconds, microseconds, etc.

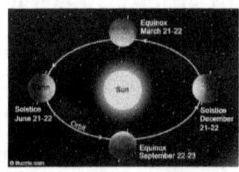 

Figure 24 – 1. Earth's Orbit and Moon's Orbit

You can also define a year in terms of a month based on the moon's orbital period of 27.3 days around the Earth. Approximately twelve orbits would be a year as show in Figure 24 – 1.

Then there is sidereal time which is a time scale based on the Earth's rate of rotation measured relative to the fixed stars rather than the Sun. In each case the Earth's or the moon's rotation is involved.

How about the universe? It is said to be about 13.7 billion years old. That statement means that during this time the Earth would have made 13.7 billion orbits around the sun. Whoa. Think about that. There is a funny fact of physics here. Before five billion years ago the Earth did not exist so what was a year 13.7 billion years ago or even just six billion years ago? There was no time unit then that could be called a year. How then can you tell the age of the universe in years?

Time Will Tell

Other Time Scales

Instead of Earth years you can try something more fundamental. For example you could count the inversions of an ammonia molecule and define a second as the time it takes for the molecule to make about 2.39×10^{10} vibrations. Multiply that by 60 to get a minute, 60 more to get an hour, 24 more to get a day, and 365 more to get a year. So a year is the time it takes for an ammonia atom in an ammonia molecule to vibrate 7.37×10^{17} times. Okay? Not so.

Another funny fact of physics is that there was a time in the growing universe when ammonia molecules, or any kinds of molecules, did not exist. How then can you use them to measure times before they were present? One answer is to try to use something that existed before molecules say atoms. The frequency of emission of excited cesium atoms has been used as a time standard. But there was a time in the history of the universe before atoms formed. So they don't work. Before atoms existed there were ions, and scientists have used mercury ion oscillations as a time standard. But there was a time before ions existed. Then the universe was made of electrons, photons, protons, neutrons, and quarks. That would take us back almost to when the universe started, the Big Bang, back to the beginning before these particles and photons existed before ions, atoms, molecules, and the Earth developed. So there is a problem.

Big Bang Time

For every problem, however, there is at least one solution. One solution to determine the age of the universe lies in the total universe. Astronomical and cosmological observations have shown that the universe (actually space) is expanding and its expansion rate is accelerating. From observations of far off galaxies we know the current rate of expansion of the universe in say light years per year. Where a light year is the distance light can travel in one Earth year. In this case you can use years to describe times like the age of the universe because you are doing

Funny Facts of Physics

this now, and the Earth and sun exist. So you can use the expansion rate of the universe to calculate its age by extrapolating backwards in time from now to when it would shrink down to a point named the Big Bang. In doing this cosmologists could use any time scale and any unit, so for convenience and familiarity they selected years.

Here's another funny fact of physics. It is impossible to mathematically shrink the universe back toward the Big Bang to a point. That point is called a singularity. We don't know what that means. It is an unknown quantity. We cannot reach it. But scientists have been able to extrapolate back to a small fraction of a second after the Big Bang. That time is called the Planck Time. For a discussion see Appendix 24. The Planck time is 10^{-43} seconds.

So scientists have extrapolated the time of the universe from the present time back to 10^{-43} seconds after the Big Bang. Why can't physicists extrapolate back further in time? The answer is because at and before the Planck time, at the singularity, the known laws, rules, and facts of physics may not apply. So in principle one can understand the universe using the laws of physics only back to that Planck time.

There is another question and a funny fact of physics. We know the time after the Big Bang to a minuscule 10^{-43} seconds out of 13.7 billion years. That's a stupendous accuracy. Yet we know the age of the universe accurately only to within ± 20 million years. How can these two vastly different accuracies be reconciled? The answer is that the Planck time is not a measured time but a calculated time. It is calculated from some constants of nature that hold at any time, early in the universe as well as now, and probably for the life of the universe. The equation for the Planck time is given in Appendix 24.

Conclusion

Various methods of estimating long times like those of the

Time Will Tell

age of the universe use Earth years even for times when the Earth did not exist. You can overcome this problem by extrapolating back in years from now. So time will tell, and you can tell time from 10^{-43} seconds after the beginning of the universe, the Big Bang, until now, almost 14 billion years later.

Funny Facts of Physics

CHAPTER 25

NO TIME BEFORE TIME

There was a time when . . . . It's time to . . . . The time will come when . . . . These are all familiar expressions you use without much thought about them. They show that time has a past. Time has a present. And time has a future. Actually you know time had a past because you remember events that happened. You are now in the present because you are conscious of your existence and sense things around you. But you can only assume time will have a future, possibly an indefinitely long future with no end to time.

Time is a part of everything you are and every act in which you engage. Your life is dictated by time. You have a time to get up, go to school or work, eat meals, and see friends. Movies and television programs are scheduled at certain times. Your smart phone shows you the time when you turn it on to text a person the time you will meet that person.

You describe yourself by your age which is how much time elapsed since you were born. A year is an important time element to you without thinking that it is the time it takes the earth to revolve around the sun as discussed in Chapter 24. Scientists predict how much time has passed since the earth was formed. It takes time for ice on a lake to melt as spring approaches. Even your cup of coffee takes time to cool before it is comfortable to drink. On and on we could talk about time. It is inescapable. We live in it. So? Did you ever wonder what it would be like to not have time?

Before Time

Well, a funny fact of physics is that there "was a time" when there was no time. The generally accepted theory of cosmology is that the universe was created with a Big Bang. There was nothing but vacuum, and from it particles, forces, and electromagnetic radiations appeared and merged, separated, expanded, and evolved with "time" to become the universe in which you now live. The universe now is made of all kinds of objects especially galaxies that contain hundreds of millions of stars. And, as time passes, these galaxies move away from each other. The universe, or space, is expanding.

The Beginning of Time

An important part of that theory is that time started at the Big Bang. Before the Big Bang time did not exist. So you ask, "When did the Big Bang happen?" A funny fact is that the question simply cannot be answered. The question has no meaning because time did not exist until the Big Bang gave birth to it.

A vacuum is nothing. No mass, no forces, no radiations, no space, and no time. That might lead you to ask, "How did all those particles and other things come into existence in a vacuum?"

The prevailing theories suggest that in a vacuum there is a fluctuating quantum field that produces virtual particles. They appear then disappear so fast that for practical purposes they do not exist, so they are in a vacuum. For more on this subject see Chapter 22. The Big Bang was an event when massive numbers of these particles suddenly created matter-antimatter pairs in large quantity. This became the universe, and that was when time started. Georges Lemaitre, a Belgian Priest and astronomer called it "A day without yesterday."

Conclusion

Time governs your life and is an inseparable part of everything. Yet time is not something that always existed. It came into being with the Big Bang. There was no time before time.

CHAPTER 26

NO SPACE OUTSIDE OF SPACE

I was at that place. I'm here now. And I intend to go there. These expressions you use without thinking much about their subtlety. They are decisions you have made, are making, or will make about your location in space. You take for granted that places to travel to and the areas all around you go on endlessly and have always been there. "After all it's just space," you say. Don't be too sure.

Space is Bounded

A funny fact of physics is that, as mentioned in Chapter 22, space has a boundary and has not existed forever. In a similar sense as for time, space did not exist before it was created at the Big Bang. There was just vacuum. Like time, space also was born from quantum fluctuations in the vacuum we call the Big Bang. Everything that exists now in our observable universe is in the space that was created at the Big Bang. And the universe is expanding, or it should be said that space is expanding. So another funny fact of physics is that when anyone talks about the universe expanding and evolving they really mean that space is expanding and the universe is evolving.

Big Bang Was Everywhere

In the case of time the question about when did the Big Bang happen has no answer because time did not exist before the Big Bang. In the case of space a question about where did the Big Bang happen has no meaning because space did not exist before the Big Bang. In that sense the Big Bang happened everywhere in what is now space.

The prevalent theory is that the Big Bang started in a point

Funny Facts of Physics

so small some call it a singularity, and the principles of physics may not apply at a singularity. Whatever, that singularity–that minuscule point–was the entirety of space at that moment when time and space began. So one can say that the Big Bang happened at the beginning of time and everywhere in space.

More Vacuum

Another funny fact of physics is that because space was conceived in the vacuum at a point and space is expanding that suggests that outside of space–outside of the known universe–there is still vacuum.

Now you might ask, "What is the difference between empty space and vacuum?" The answer is that vacuum is nothing. Space differs from vacuum because it cannot be empty. It exists, and the universe that occupies space is made of masses, forces, fields, and radiations. There may be voids between some of the masses, but space is there, and there is some density of matter and radiation in space. Vacuum, on the other hand, is truly nothing (except for virtual particles). No mass. No forces. No fields. No radiations. No space. No time.

Conclusion

Space was created at the Big Bang from a vacuum. Our universe occupies whatever space exists, and there is nothing beyond it but vacuum. So you cannot and never will be able to see anything beyond the space in which our universe exists. There may be multiple universes that might or might not be like the one you are in, but like what lies outside of our universe, they also would never be seen.

CHAPTER 27

IT'S ALL IN THE DARK

Big Things
Earth is big. It has a mass of 5.97 x 10^{24} kilograms. Our sun contains a mass of 1.99 x 10^{30} kg. The Milky Way galaxy distributes 6 x 10^{42} kg of mass. That's a million trillion times the mass of the earth. Now we're getting to the really big stuff. How about the entire universe? The observed universe, made up of billions of galaxies, dust, stars, gases, nebulae, and whatever we see, contains an estimated 10^{80} particles and has an estimated mass of 10^{53} kg. That's a number with 53 zeros after it. It's no wonder why the "explosion" that current theories say created the universe is called the BIG BANG.

About Galaxies
From all these objects let's first take a look at galaxies. Typically their stars, planets, and all else in them whirl around the centers of the galaxies, which probably harbor black holes. Everything spins so fast that, according to the laws of physics, galaxies should fly apart. But galaxies don't come apart because it has formerly been concluded, that the gravitational attraction of all the mass in a galaxy holds things together. That's a great idea that was thought true for a long time before the masses of galaxies were truly determined.

Too Little Mass
A funny fact of physics is that when all the mass of any galaxy goes into the calculation about holding the galaxy together by gravitational attraction, that mass is only about 5 percent of what is needed. Galaxies do not fly apart so there has to be more matter in them than can be seen. Physicists have called this

Funny Facts of Physics

unseen 95 percent of the matter in galaxies and in the universe Dark Matter, and they are searching really hard to find it.

Too Much Expansion

Now if we think of not just one galaxy but all the galaxies in the universe together, we know they are moving apart because space is expanding. And its expansion is accelerating. Another funny fact of physics is that the acceleration of space expansion is greater than what is calculated by Einstein's theory which has been justified in all experiments and other observations. There seems to be some form of energy causing this acceleration. The potential and kinetic energy available to space, as we know it, is only about 30 percent of the energy needed to cause the acceleration of the expansion of space. This additional 70 percent of unknown energy has not been discovered, so like its counterpart, Dark Matter, Physicists call it Dark Energy.

When Einstein developed his theory of general relativity, his equations predicted that the universe would expand. Observations at that time determined that the universe was not expanding. Einstein firmly believed that theory should match observations so he added a "cosmological constant" to his equations to keep his description of space static.

Later, astronomers discovered that the universe was expanding, so Einstein's constant was not needed. A funny fact of physics is that cosmologists then used the idea of a cosmological constant to explain not the restriction on an expanding universe, but the accelerating expansion of the universe.

Conclusion

For hundreds of years physicists have achieved an astonishing understanding of our physical universe. They have collected immense amounts of data, and physicists have produced numerous working speculations and developed many substantiated theories about how our world works. Now they

It's All In The Dark

discover that they have been talking and writing about and experimenting on only 5% of the matter and 30% of the types of energy that exist in the universe. Now that is a funny fact of physics.

Funny Facts of Physics

CHAPTER 28

IT MUST BE JELLY

There was once a popular song titled "It Must be Jelly Cuz Jam Don't Shake Like That." A band named The Hipp Cats were the first to record it on August 13, 1938. It became even more popular after the Glenn Miller orchestra recorded it on July 15, 1942. It became a favorite song, and young people loved dancing to its rhythm and beat. Whatever the item is that the song refers to, at least you know how jelly shakes. Well really jelly and Jello vibrate, support wave motions, and ripple. Vibrations, waves, and ripples pretty much describe the same action so I will use them in this chapter interchangeably.

You see many things about you that vibrate or move in waves or ripples. Water ripples after a stone is dropped in. Drum heads, cloud formations, strings on musical instruments, a baseball bat, a tennis racquet, or golf club after either strikes a ball, and wings on jet planes in flight show similar motions. They are caused by some stress on the medium. There are many more illustrations you could visualize if you stop and think about them for a while. You are surrounded by vibrating things and wave and rippling motion. You see so much of it you take it for granted.

Let's go back to the case of Jelly as an example. An engineer or scientist would call jelly an elastic material. If you shake it, push on it, or squeeze it, the jelly will vibrate. Elastic waves ripple through it. Jelly and all these other mentioned vibrating objects are physical materials. The word physical means they have mass and spatial dimensions. You can see, feel, weigh, and measure their size. They are what you would call real objects. Now you may ask the question, "Are there entities less real, less

Funny Facts of Physics

physical, that can vibrate and ripple?"
Space

What about space and the vacuum? Can they ripple? You can't see, weigh, or measure the dimensions of space or the vacuum. Are space and the vacuum real physical entities or are they some kinds of nebulous entities? I discussed in Chapter 22 the virtual vacuum, and in Chapter 26 I wrote about some features of space. I mentioned that space was created from the vacuum at the Big Bang and that space is expanding at an accelerating rate. So space exists, though it does not have mass, therefor it too should be able to support vibrations or ripples in its structure.

Ripples in Space Predicted

The idea of vibrations or ripples in space was not thought about or looked for over a long time while physicists and astronomers searched out, discovered, and explained many other physical and astronomical phenomena including the theory of space.

A funny fact of physics is that In 1916 Einstein, using conclusions from his Theory of General Relativity, predicted that waves in space-time, called gravitational waves, should exist. Neither he nor other persons did much theoretical or experimental research on this conclusion. Not until 1936, while doing further research on his relativity theory, Einstein found reason to repudiate his prediction about the existence of gravitational waves. His conclusion then was that they could not exist. But a mistake discovered in this new research, when corrected, once again confirmed his previous prediction that gravitational waves should exist. Einstein stated, however, that the waves would probably never be measured.

Ripples in Space Observed

For almost 100 years Einstein's idea of gravitational waves had not been confirmed, but in 2015 astronomers startled the world of science when they observed gravitational waves referred to as ripples in space-time. They postulated that the

It Must Be Jelly

waves they measured were caused by the merging of two black holes with masses of 29 and 36 times that of the sun. These waves were observed by an experiment called LIGO, after the Laser Interferometer Gravitational-Wave Observatory. According to Dr. David Reitze, executive director of LIGO, "The gravitational waves stretched and compressed space around Earth 'like Jell-O.'"

In August 2017 scientists at LIGO using two detectors and a detector from the Advanced Virgo Detector saw more gravitational waves . They were from the merger of two black holes of 30.5 and 25.3 times the mass of the sun.

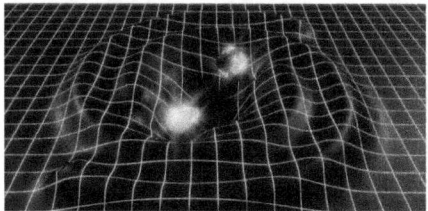

Figure 28 – 1. Gravitational Waves

Figure 28 – 1 is a simulation of space time as a two dimensional elastic medium. Two black holes have collided near the center of the picture and it shows how they send out a shock that causes waves, or ripples, to propagate in the elastic medium of space-time.

Conclusion

All physical mediums can support vibrations or waves or ripples if they are stressed by a force. Some mediums are more rigid than others, but to some degree they all can vibrate. This reaction is true even for space-time in which gravitational waves predicted by Einstein may occur when massive objects like black holes collide.

Funny Facts of Physics

CHAPTER 29

137 – THE UNIVERSE'S PIN CODE

Humans have long wondered about the mysteries of the universe, its size and structure, how it came into being, what is its destiny, and of course what is humankind's status and role in the universe. Why is the universe here? Why are humans here? Is life on Earth the only life in the universe? Answers to the "how" and "what" questions are being explored, being answered, and may some day be more completely answered. Answers to the "why" questions are not in the realm of science. They are the provinces of metaphysics, philosophy, and theology. As to life elsewhere in the universe, many persons are studying, exploring, and devising schemes to learn if life is unique to Earth, occurs occasionally elsewhere, or is ubiquitous in the cosmos.

The Large Scale

Despite the lack of answers to many questions, scientists have determined that on the large scale the universe is made of stars, planets, galaxies, nebulae, black holes, and a variety of objects changing their form from one to another. Then too, scientists suspect the existence of and are looking for dark matter and dark energy, things that have never been detected but are suspected to exist. Some of this issue is discussed in Chapter 27.

The Small Scale

On the microscopic scale scientists have discovered a system to the universe. Everything, even those things on the large scale, seems to be made of very small fundamental particles like photons, molecules, atoms, electrons, protons, neutrons, quarks, and a slew of more esoteric particles. Everything that is in our everyday world and in the observed universe is made of those same fundamental particles. The not yet observed dark energy and

Funny Facts of Physics

dark matter may be made of these same or possibly different small particles.

Models of the Universe

What is the basis of all this matter whose existence scientists say started about 13.7 billion years ago and has evolved into what it is today? Could something other than our observed universe have happened? Is the observed universe the only one?

Scientists like Niels Bohr and Arnold Sommerfeld derived equations that explained the structure and actions of atoms and their nuclei. James Clerk Maxwell's theory, given in a set of four equations, explained electromagnetism. Albert Einstein gave the world a set of equations that describe nature when things move near the speed of light or are in large gravitational fields. Werner Heisenberg, Erwin Schrodinger, Louis de Broglie, Paul Dirac, Albert Einstein, and other scientists put the world of the small, like atoms and their components, on a mathematical footing. A host of researchers like Murray Gell Mann, Tsung-Dao Lee, Chen-Ning Yang, Richard Feynman, and other physicists developed theories that culminated in what is now called the Standard Model, Quantum Electrodynamics, and Quantum Field Theory that explain most of the tiny subatomic particles in the world and the evolution of the universe from those particles. Recently the basis of the masses of all particles may be explained by the discovery in 2012 of the Higgs boson named after Peter Higgs who predicted the existence of the particle in 1964. Recently in 2016 two experiments appear to have seen gravitational waves that Albert Einstein predicted one hundred years ago. See Chapter 28.

What a tremendous catalog of knowledge of the universe physicists have achieved from tiny quarks inside protons and neutrons to the massive Higgs Boson to the colliding of black holes and ripples in space-time.

Let's look at these ideas from three angles. One is Electromagnetism that represents older physics and the large

137 – The Universe's PIN Code

scale, mostly everyday, world. It involves the electronic charge on an electron, a constant of nature, given by the symbol, e. The second is Relativity, a newcomer in 1905, based on the speed of light in a vacuum, a constant given by the symbol, c. The third is Quantum Theory, the newest and still developing theory that deals with quanta whose size and energy are given in terms of another constant quantity, Plank's Constant given by the symbol, h. Speculations and theories of biology also are based on laws of physics which incorporate those three areas of physics and their constants.

Fine Structure Constant

In all theories of nature these three constant quantities, e, c, and h, appear in some manner, especially as the combination e^2/hc called the Fine Structure Constant and given the symbol the Greek letter, α. This constant has no dimension. It's just a number. Its reciprocal, $1/\alpha$, has the value about 137.

A funny fact of physics is that if the numerical value of α differed by a tiny amount the equations of electromagnetism, relativity, quantum theory, and even the fundamentals of cosmology and biology would not describe nature. In other words, if α had a slightly different value, atoms, stars, the universe would not exist and you would not be here to observe them. It is as though the value of α was arranged so that humans could exist. This proposal has the name Anthropic Cosmological Principle.

Richard Feynman, world-famous physicist, and Nobel Laureate who explained the reason for the Challenger Shuttle disaster, called $1/\alpha = 137$ *"a magic number"* and its value *"one of the greatest mysteries of physics: a magic number that comes to us with no understanding by man. You might say the 'hand of God' wrote that number."*

Frank Close, Professor of Physics at the University of Oxford who has received many professional awards for his research and writing stated, *"Alpha sets the scale of nature – the size of atoms and all things made of them, the intensity and colors*

Funny Facts of Physics

of light, the strength of magnetism, and the metabolic rate of life itself. It controls everything that we see. In 137, apparently, science had found Nature's PIN Code."

Conclusion

The funny fact of physics is that your existence, your being here to read these words depends on the number $1/\alpha = 137$ being just that and not some even slightly different number. Your existence depends on the universe having the correct PIN Code–137.

PART SIX

FUTURE PHYSICS

Funny Facts of Physics

CHAPTER 30

THE END OF PHYSICS

You may wonder why I place a chapter about the end of physics in the section on the future of physics. The reason this chapter is here is to inform you that there have been numerous times in the history of physics (as in other sciences also) when the practitioners of the science thought seriously that they knew all that could be known about the subject, that the end of their science was at hand. But despite their esteem and recognition by their peers as experts, and consensus among colleagues who accepted and agreed with their opinions, they were wrong. In most instances not only was the end of their science not near, but soon after their pronouncements, their science began a rebirth and explosion of new facts and theories to explain them. In many cases those same "end of physics" proclaimers contributed to the advance of physics after their statements of doom. For example the following quotes by renowned physicists illustrate some statements about the proposed end of physics.

1894 – "The more important fundamental laws and facts of physical science have all been discovered." – Albert Michelson.

Albert Michelson was awarded numerous prizes for his research and won a Nobel prize for measuring the speed of light and demolishing the speculation about a luminous aether. This work, contrary to his quote about everything having been discovered, contributed to the then new and revolutionary theory of relativity.

Funny Facts of Physics

1927 – "Physics as we know it, will be over in six months." – Max Born.

Max Born won a Nobel prize for his work on statistical interpretation of the quantum mechanical wave function, and he made numerous contributions to the development of quantum theory including the matrix formulation of the theory. His research actually developed new ideas about the atomic world that are still being explored almost nine decades later contradicting his quote about physics ending in six months.

Theory of Everything

Complimentary to the idea that physics had reached its end because physicists knew all there was to know about the physical world is the idea that all the knowledge we do have, and all the separate theories that organize that knowledge, can be described by a single theory. This grand scheme would be a theory of everything.

Current Theories

Today the two dominant theories that explain much of the physical world are general relativity and quantum mechanics with its extensions of field theory and quantum electrodynamics. But they describe different regions of the universe. Relativity deals with velocities near the speed of light and very large masses. Quantum mechanics is a theory of the very small world of atomic and subatomic particles. Melding those two theories together would bring us close to a single theory that describes the universe from the smallest to the largest scales. There was a time when some prominent physicists thought they could achieve that goal. For example:

1980 – "There is a 50 percent chance that we would find a complete unified theory of everything by the end of the century." – Stephen Hawking.

The End Of Physics

Stephen Hawking is well known for his popularization of time, quantum theory, and cosmology. He is considered by many physicists as one of the brightest scientists of our era. He has not received a Nobel prize, but many of his contemporaries feel that he deserves one.

Despite Hawking's quote, for more than the next three decades after he made that projection, his own research has added much to the knowledge of cosmology and quantum theory. Continuing efforts by Hawking and many other physicists, however, have not yet found the unified theory he and other physicists proposed.

No Theory Yet

A funny fact of physics is that these expressions by leading physicists that declared an end to physics or the discovery of a theory of everything were offered by three of the most eminent physicists who advanced our knowledge of the physical world. There were many more physicists who echoed their sentiments, and there may have been a consensus among physicists that there could be an end to physics or a theory of everything. And more astonishing is that each of these persons, and others of their persuasion, contributed much more to the development of physics long after they made their statements.

There Is Doubt

These opinions about the end of physics embrace almost a century of thought about physics and show that the idea of an end to physics was conceivable, but so far not true. Currently most physicists have recognized that there are mysteries of the universe still unseen and uncovered and unexplained. They understand now the irrationality of making statements about the end of physics and wisely refrain from issuing such opinions. Many researchers also feel that a theory of everything is unattainable because there will always be some newly discovered physics that does not fit into that theory.

A theme of physics today is that we can explore and

Funny Facts of Physics

question the physical world almost endlessly on both the small and the large scale and continue to find new phenomena that raise more questions than our explorations provide answers.

Conclusion

Physicists have discovered an endless succession of wonders in the physical world and developed a number of theories to explain them. And they have found many unexplained phenomena like dark matter and dark energy. Speculations abound and theories are being developed, but the theories like quantum theory and relativity are diverse and not comprehensive. The end of physics and a theory of everything are, like the Earth's horizon, forever moving targets.

CHAPTER 31

TIE IT ALL TOGETHER

No end of physics is in sight, and the search for a theory of everything continues. But that all-inclusive theory remains elusive. If a new theory of everything can be found, it must contain those theories that have worked well in their respective domains such as Newtonian mechanics, Maxwell's electromagnetism, thermodynamics, general relativity, quantum theory, quantum electrodynamics, field theories, and others.

Current Theories

James Clerk Maxwell's theory, embodied in his four equations, explains electromagnetic phenomena from magnetic phenomena to static electric charges to electric currents to electromagnetic radiations. Sir Isaac Newton's mechanics explains many observed facts and predicts motions of objects from the tossing of a ball to those of sending men to the moon and launching probes to outer space. Thermodynamics explains the properties and actions of matter under the influences of temperature and pressure and the limits on exchanges of various forms of energy. Albert Einstein's Special Theory of relativity explains observed changes in properties of objects and changes in time intervals of things moving at constant relativistic speeds. His General Theory of Relativity describes the influence of large masses on everything even light and how mass affects space and space affects mass. These theories of Einstein include those of Newton in the low velocity realm. Quantum theory, and its extensions and refinements, explain the observed phenomena of very small objects like electrons and nuclear particles.

Physics has identified electromagnetic, gravitational, strong, and weak forces that explain electromagnetic actions,

Funny Facts of Physics

gravitational actions, the binding of nuclear particles, and the process of actions like radioactivity. Physicists have shown how at one time some of these four forces were joined. In continuing attempts to demonstrate this they have reduced the number of independent forces to three – gravitation, strong, and electroweak. And theories now join the strong force to the electroweak force

The two complementary theories of physics, quantum theory and relativity, have led to a working model of the creation of the universe, how it evolved, and the roles of the many small particles and forces that make the universe of large and small scales what it is today. These theories are called Quantum Electrodynamics, Quantum Field Theory, and the Standard Model. They work well in their domains of applicability, but they have flaws and cannot explain everything or be combined into one Theory of Everything.

Not All Theories Work

One problem is that there are phenomena that cannot be explained by either general relativity or quantum theory, and each theory has a domain in which it can be applied. General relativity applies to large and massive objects. Quantum theory applies to very small objects. A prime example of where the two theories are incompatible is a black hole. It is a very massive object so some features obey general relativity theory. On the other hand it is a very tiny object, maybe even a minuscule point becoming a singularity, and here quantum theory should apply. But the two theories cannot be used together to explain all the physics of a black hole. The same problems exist in trying to explain the earliest moments of the Big Bang. What is needed is a theory of "quantum gravity" that would combine the features of quantum theory and relativity. Such an accomplishment would take a huge step toward a theory of everything.

That desire to reduce all theories to just one theory has been a goal of many of the leading physicists for more than a century. Einstein, for example, devoted the latter years of his life

Tie It All Together

to that endeavor without success. No one has been able to tie all the theories together using known and familiar physics. Many researchers are working hard developing new physical ideas trying to unify the theories into a theory of everything. They are trying to tie all the theories together. A funny fact of physics is that they are trying to tie the theories together with string.

Try Using String

This area of research is called String Theory. It developed from the idea that elementary particles do not act like little billiard balls or even waves as discussed in Chapter 15, but instead they act like tiny vibrating strings as suggested by the sketch in Figure 31 – 1. A variety of string theories were developed, and a Superstring Theory is being attempted to combine the separate string theories.

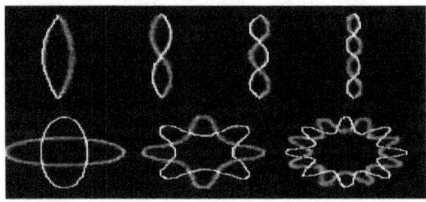

Figure 31 – 1. Schematic Drawing of Strings

A string that vibrates in different ways suggests that the different vibrations represent different fundamental particles. If a string vibrates in one mode, it would act like an electron. If it vibrates in a different mode, it might act like a proton. A funny fact of String theory is that it requires, in addition to the three spatial dimensions and one time dimension, possibly six more spatial dimensions. If string theory is correct that means space-time might have ten dimensions not four. The arguments given for not seeing these six other dimensions are that they are tiny and hidden in side the elementary particles, i.e., strings.

These other dimensions would appear to us only at exceedingly high energies or temperatures. These are energies

Funny Facts of Physics

much higher than we can produce now even in particle accelerators. So for now there is no experimental evidence for strings.

Conclusion

You should ask, "Can those extra dimensions ever exhibit themselves? Can we find physical evidence that strings exist? If they do exist will string theory produce that long sought goal of a Theory of Everything?" And finally, "Will strings tie it all together?"

CHAPTER 32

PHYSICS HAS MANY FUTURES

The last and most comprehensive funny fact of physics is that physics not only has a future, but that it has many futures.

Physics is a study of the physical world with the goals of understanding how the physical world works and to understand the universe and the part life, especially human life, plays in it. Other goals of physics are to use the knowledge gained to improve the pleasure, health, and longevity of humans and to give people a more peaceful, secure, and happy life.

For a few hundred years physics has been progressing in the goals of understanding the universe. Physicists have observed the universe from the very small world inside nuclei to the very large world of space and galaxies. They developed theories that, in most cases explain things satisfactorily enough to say that we know how many things work including the birth and evolution of the universe.

Still they use different theories for different realms of force, mass, size, time, and speed and they would like to combine these theories into a single theory. In addition, there is more work to do in traditional areas. Beyond current ideas and quests, physicists are pushing the frontier of knowledge to other physical areas needing explanation. Physics is still an open discipline and has a future in developing theories about how everything, not just the traditional physical subject areas, but all of life, works.

Now physicists are increasingly turning their attention to other areas not considered "physical" but which really are. This conclusion is so because everything is made of molecules and atoms and the fields and forces that hold things together. This

Funny Facts of Physics

surely has been, is, and always will be the realm of physics. Some of these new areas to be explored by the methods, concepts, and laws of physics, given here in arbitrary order, are:

1. Traditional Physics

Many well-known phenomena are being reexamined with more precise measurements to determine if the current theories hold or need improvement. Laboratory staffs are busy exploring the properties of materials under extreme temperatures and pressures like those found in stars, for example. The land of the small, submicron size materials, are being studied and are showing unique features and applications.

2. Cosmology

A search for what makes up most of the universe yet has not been observed. Dark matter and dark energy will produce new physics. Recently observed gravitational waves, which Einstein first suggested existed in 1916, will open a new way of looking at, studying, and explaining the universe.

3. A Theory of Everything

Today there are various theories each of which explains a different area of the physical world. Worldwide, physicists are working independently and collaborating on showing that the various theories like relativity and quantum theory can be joined to produce one theory that explains all observed phenomena.

4. Communication and Computing

Physicists have been instrumental in developing the solid state and atomic physics that are fundamental to most of our electronic devices. Now they are working on moving to smaller scales and true quantum computing.

5. Economic and Social Phenomena

Many methods of statistical mechanics, entropy, and fluid motions are being applied to the study of economics, business, financial markets, and social interactions.

6. Medicine

Physics has given the field of medicine radioactive

Epilogue

treatments and ultrasound, optical, x ray, and magnetic imaging. Neutron and proton and other imaging procedures are being expanded. The techniques and philosophy of physics are being used in various medical studies, particularly in the attack on cancer.

7. Searching for Extraterrestrial Life.

This search has been under way for many decades with no evidence that life exists elsewhere in the known universe. The search will continue with more accurate and precise detectors and with new and novel methods of detecting life. In addition to traditional ground-based telescopes and sensors Earth orbiting satellites will be used. Space vehicles that have orbited and landed on other planets in our solar system are looking for life forms or the chemical elements needed for life. The search for extraterrestrial life has different approaches in including looking for chemical signatures, unintelligent life, and intelligent life with advanced technology. These efforts will continue and expand.

8. Treating, Curing, and Preventing Diseases and extending life.

Physics has been involved in medical science through x-ray and other imaging methods such as MRI (Magnetic Resonance Imaging). Physicists have developed radiation techniques to conquer malignant diseases. Lasers are used for diagnosis, treatment, and prevention of health problems. Physicists are applying physical ideas and methodologies to chemistry and biology.

9. Understanding the Origin of Life.

Physicists have been applying the laws of thermodynamics, chaos theory, and other emergent phenomena to try to understand what life is and how it started. This is one of the most difficult tasks because scientists of all disciplines have not agreed on what life is. Physics will help establish an accepted definition.

Funny Facts of Physics

10. Difference Between the Brain and the Mind.

The brain is made of atoms, fields, and forces. Speculations are that the mind is different from the brain. Therefor the mind is not a physical entity. If this idea is correct then the mind cannot be studied by the methods of physics, chemistry, or biology. Physics certainly will be involved in solving this dilemma.

11. Puzzling Over Consciousness.

Related to the relationship between the brain and the mind is the issue of consciousness. Humans are aware of their existence and mortality. We recall the past, know the present, and speculate on the future. Most persons believe that only humans have this sense, and this is one characteristic that distinguishes humans from other life forms. Physicists will be involved in understanding if consciousness is a physical entity or something nebulous.

Conclusion

In addition to these few ideas I sketched rather casually and incompletely there are other ideas known and some unknown that will emerge as research continues and new phenomena are discovered. The funny fact is that there will always be surprises in science. Physics has a long, varied, and exciting future in which to expand knowledge of our world in many diverse areas and advance the goals of the science of physics.

EPILOGUE

PHYSICS
METAPHYSICS
PSEUDOPHYSICS

Funny Facts of Physics

EPILOGUE

PHYSICS, METAPHYSICS, PSEUDOPHYSICS

In this book I discussed funny, by that I mean strange, facts of physics. Though you might consider them strange and unexpected from your everyday experiences, all the phenomena mentioned have explanations within the methodology, rules, laws, and theories of physics that have been tested over many years and have never been proven false. Physics is the study of physical phenomena using observation, measurement, experimentation, and mathematical modeling leading to theories, and then attempts to falsify the theories. There are two other fields of activity that are somewhat related to physics, and I compare them below with physics.

PHYSICS

Physics is a science. It explains observed and predicted phenomena using empirical evidence, numerical measurements, facts, and logic. With these tools, physicists make a hypothesis then test it against evidence and calculations. If the hypothesis agrees with the physical evidence and is duplicated by other physicists the hypothesis is accepted as true and is called a theory. Yet no matter how well a phenomenon may be explained by a theory, physicists can never be certain of their conclusions so they try to demonstrate that the conclusions are false. If a falsification is successful then the theory is deemed wrong and has to be modified, improved, changed, or discarded and replaced. If a theory cannot be falsified, it is regarded as the best explanation of a phenomenon, and the theory is accepted as the best explanation available. In that manner physics is a continuing search for truth by trying to prove that its conclusions are false. This approach to understanding our universe is unique to science.

Funny Facts of Physics

Despite being able to explain nature and the various parts of the universe, physics can only answer questions about how the world works, how it came into existence, and predict its future. Physics cannot answer questions about why the universe works as it does or why it exists. Also outside the realm of physics are questions about why humans are here or the existence of God.

METAPHYSICS

Another discipline related to but which differs significantly from physics is metaphysics. The prefix "meta" means *after, along with, beyond, among, behind*. So metaphysics goes outside of and beyond physics. It is an adjunct to physics. Metaphysics is an approach to understanding the world that differs from that of physics in that metaphysics does not use the empirical, numerical, and mathematical methods of physics. And it does not use falsification of its ideas as a fundamental approach to understanding the world. Therefor it is not a science.

Though metaphysics does not employ the procedure of falsification, it does use logic and selection of material evidence to prove desired points of view about why things are as they are. It is more in the realm of philosophy and religion than science. Though it is not a science, metaphysics is a rational field of study, and is a reasonable adjunct to physics because it raises questions that physics cannot answer. It attempts to answer those "why" questions not answerable by physics such as, Why is there a universe? Why are we here? Is there a God?

PSEUDOPHYSICS

There is unfortunately a third kind of "physics" called pseudophysics. The prefix "pseudo" means *false, fake, lying, sham*. Pseudophysics appears to be informative and harmless, but it is a danger to science and society. What is it, and why is it a danger?

You saw that many facts of physics are funny. It is also funny, but in an unfortunate way, that many facts of physics and their accompanying theories have been changed, denied,

Epilogue

misinterpreted, and distorted. This distortion of physical phenomena and theories is pseudophysics. The practitioners of pseudophysics use facts and theories in ways not intended by physicists. Pseudophysics appropriates and misuses terms and concepts used and understood in physics and which explain and correctly predict observations of the world. That misuse of physics makes the activity of pseudophysics not just funny but dangerous. People exposed to this charlatanism think they are being given real physics but are misled. They are told the world works differently from the way it does, and these pseudo physicists can induce persons with little knowledge of physics to make dangerous technical, social, and political decisions.

This distortion of reality is the nature of pseodophysics. So it is an untruthful therefor a dangerous, activity. Practitioners of pseudophysics lead gullible people to believe what is not true, demean the theories and explanations of real physics, and influence social and political decisions including political financial support for pseudophysics rather than for responsible physics.

Pseudophysics is not a science. It does not rely on facts and logic. In refutes many facts and logical conclusions. It attempts to verify not falsify its assertions. It relies on faith, mysticism, superficial beliefs, personal desires, coincidence, and anecdotal events rather than on observations, facts, and logic. What are some of these categories of pseudo physics?

Energy

A word those pseudoscience enthusiasts have adopted from physics is energy. The word energy has been given many meanings though the users have no idea of what energy is. Karmic energy and qi, supposedly a life force that flows through the body, are two examples of "energies." And by extrapolating from Einstein's equation $E = mc^2$ they equate their mystic energies to the mass of one's body giving the body some kind of aural "energy" that if focused can cause dramatic changes. This interpretation of Einstein's equation shows an utter lack of understanding of the equation, its meaning, and its limits. See Chapter 4 for a discussion of energy.

Pure Energy

There is no instance in nature in which mass transforms into "pure energy" (or vice versa) without some material particles or radiation carrying that energy. Pure energy is a meaningless phrase. There is no such thing "pure energy." Energy is a property or state of a material or radiation and is a measure of doing work. It has to have a form such as nuclear, mechanical, chemical, electrical, thermal, or radiative. For more details see Chapter 4.

Soul-Matter Equivalence

Some misled persons conclude that because Einstein's equation shows an equivalence between energy and mass there is an equivalence between soul and matter. They imply that energy and soul are similar. There is physical evidence for energy in its proper physical sense, but there is no evidence for soul.

Body Energy

Some pseudo healing practitioners claim that energy channels exist in the human body. They assert that energy flows through these channels and influences a person's health and mental and emotional states. These ideas include the concept of energy, but without an understanding of energy. The first thing wrong is that they conceive of this supposed energy that flows through the body as a material substance. The second problem is that they do not define and identify the channels in the body. In fact, no examiners of living or dead bodies have found these assumed channels or this supposed energy. The energy that they claim exists in body channels is nonsense because energy is not a substance. See Chapter 4 for a discussion of the meaning of energy.

Creationism

The ideas of creationism have a long history that evolved from ancient religious dogma and incomplete scientific

Epilogue

knowledge. Even today many, mostly less well educated persons, believe in creationism and that Darwinian evolution is a hoax. They are encouraged by proponents of this doctrine which asserts three things. First, the universe, including especially the earth and all things on the earth including humans, was created in the form that it and its contents are in now by an omnipotent God 6,000 years ago. Second, the story of Genesis in the Bible is factual. Third, God creates a soul for every human at birth.

In this type of pseudophysics the practitioners resort to faith, religion, and personal desires rather than facts and physical laws. Their ideas are accepted by gullible persons and promoted despite evidence disputing their claims. There is evidence of animal, pre-human, and human fossils from many more than 6,000 years ago. Radioactive dating of components of the earth and astronomical observations exhibit galaxy lifetimes of ten billion years, the Earth's life of five billion years, and homo sapiens existing for 100,000 years.

Scientology

Proponents of scientology promote the idea that man consists of body, mind, and soul, but is spiritual with properties beyond those observed. It promotes mythical-like personages and previous lives. Though it claims to be science based, none of its ideas can be observed, experimented with, and proven to be either true or false. And, as different from science, it makes no attempt at falsification. It is considered by many persons a religious sect.

Astrology

Astrology is one of possibly the oldest of the pseudosciences. Some people still believe in it despite continuing evidence that its suggestions are arbitrary. There are no correlations between stars, planets, and other bodies in the universe and people. Its predictions are not accurate. When astrological predictions agree with events, they are purely coincidental, but practitioners proclaim these coincidences as evidence of the truth of astrology. Astrology is not a science. It

Funny Facts of Physics

does not use facts and logic. It is now offered in newspapers usually with the comics and puzzle sections. That's reasonable because astrology is no more than entertainment.

Telekinesis

There are frauds who profess to be able to exert their mind over matter as in bending of spoons or predicting the numbers on cards in a deck by thinking of them. In all instances these events have been shown to be fakes. The brain emits electromagnetic radiations when it is active, but they are of very low intensity. Very delicate and precise instruments placed close to a person's head are needed to observe them. They do not radiate nearly enough energy to change the shape of matter or influence it in any significant way.

Conscious World

Some misguided persons still think that "nature abhors a vacuum" or that "water wants to flow downhill." These expressions propose that inanimate bodies have a mind or consciousness; they have wants and needs. In the experiments with electrons or photons passing through slits some non scientists suggest that the projectiles are conscious and "know" when a slit is open or closed. See Chapter 15. The electrons would know to go to the proper slit to create the pattern that is observed, both in the single slit and double slit experiment. Such ideas have no basis in fact. Sticks and stones do not have minds or consciousness, and neither do electrons nor other atomic or nuclear particles. This idea of matter having consciousness is called panpsychism. It proposes that all matter, organic or inorganic and regardless of form has consciousness. There is no scientific or other evidence to support that proposal.

All Is Relative

Another example of pseudophysics is the notion that because Special and General Relativity are deemed correct, the ideas embraced by the theories must apply to all everyday aspects of life. They apply this doctrine to physical objects regardless of

Epilogue

their moving at non-relativistic speeds at which everyday life proceeds. Adherents of this idea assume that because relativity predicts that massive objects like galaxies can bend light, the same can be done with small objects. And they even apply the relativity concept to nonphysical things. "Everything is relative," they say. Thus there are no lies only relative declarations. Good and bad are relative to the events and people to whom they apply. And because time and space are not fixed, not absolute, nothing in life is absolute.

Conclusion

These are a few examples of the ways in which the true science of physics can be distorted to appear reasonable, but they have no factual justification. In this book I discussed funny, by that I mean strange, facts of physics. Though you might consider them strange and unexpected from your everyday experiences, all the phenomena mentioned have explanations within the methodology, rules, laws, and theories of physics that have been tested over many years and have never been proven false. Physics is the study of physical materials and phenomena using observation, measurement, experimentation, and mathematical models to develop theories that explain observations.

Funny Facts of Physics / Epilogue

Epilogue

APPENDICES

Funny Facts of Physics

APPENDIX P

LOGARITHM SCALE AND SCIENTIFIC NOTATION

The scientific notation for numbers is also called "powers of ten," "exponential," "orders of magnitude," or "log scale." Large numbers, such as 234,000,000 or 0.0000000234, are difficult to work with and to list in tables. You have to count the zeros to know if the first number is a million, or a hundred million or what. In the second case is it millionth or ten millionth or what?

These kinds of numbers can be expressed in a more compact form. Write them with one digit to the left of a decimal point. Then count the rest of the digits and zeros and write that number of digits as a power of ten (an exponent of ten). Then multiply that to the first part. For example, the number 1,000 is written 1×10^3 or just 10^3. For a number less than 1, such as 0.001, count in the negative direction and write it as 1×10^{-3} or 10^{-3}.

Examples:
10^6 = one million
10^7 = ten million
10^8 = one hundred million
10^9 = billion.

Take the first number in the first sentence. It is written as 2.34×10^8. Now you know from the number 8 there are eight digits, including zeros, after the 2. So the number is 2.34 times one hundred million. The second number is written as 2.34×10^{-8}. Logarithms condense these numbers more. A log of a number is the power of ten associated with it.

In log terms a few examples are:
$\log 1 = \log 10^0 = 0 \quad \log 100,0\,00 = \log 10^5 = 5$

Funny Facts of Physics

$\log 10 = \log 10^1 = 1$ $\log 100,000,000 = \log 10^8 = 8$

On a logarithm scale instead of using large numbers like 100000 and 10000000 one can use their logs which are 5 and 7. You will see log graphs later. The same ritual applies for negative exponentials. For example

$\log 0.001 = \log 10^{-3} = -3$
$\log 0.000001 = \log 10^{-6} = -6.$

APPENDIX 4

ENERGY DOES NOT EXIST

Definition of Mechanical Work

Work, W, is the product of two quantities, a net force, F, acting on a body of mass, M, and the distance, d, through which the body moves in the direction of the net force. This definition includes curvilinear as well as linear motion. Work done during linear motion is mathematically expressed as
$$W = F \times d.$$
The work done during curvilinear motion is
$$W = F \times r \times \theta$$
where r is the distance to the force from the center about which the body rotates and θ is the angle through which the body rotates.

Definition of Electrical Work

Using the definition $W = F \times d$ and the force on an electric charge q by an electric field E given by
$$F = q \times E,$$
then
$$W = q \times E \times d.$$

Definition of Energy

Using these definitions of work, a definition of energy, that permits you to make numerical calculations of the use of energy is:

> *Energy is a concept that describes the physical state of a material or system that allows it to perform work on or have work done by that material or system or causes a temperature change in the material or system.*

Funny Facts of Physics

APPENDIX 7

HOW TO LOSE WEIGHT

Weight is a force. As a mathematical equation, weight W of a person is given by
$$W = G m M / r^2$$
in which $G = 6.67 \times 10^{-11}$ N-m^2/kg^2 is Newton's gravitational constant, m is the mass of a person, M is the mass of the Earth, and r is the distance from the center of the Earth to the center of mass of the person. The letter N is for Newton which is a unit of force.

The gravitational force between the Earth and an object on the surface of the Earth with a mass, m, is given by
$$F = G m M/R^2$$
in which Newton's gravitational constant $G = 6.67 \times 10^{-11}$ Newton-meter^2/kg^2, M is the mass of the Earth, and R is the radius of the Earth. For an object above the Earth the equation is
$$F = G m M/ (R+r_m)^2$$
and for an object inside the Earth the equation is

$$F = G m (4/3)\pi \rho (R-r_m)$$

where ρ is the density of the Earth, R is the radius of the Earth, r_m is the distance from the surface of the Earth to the object. This gravitational force is what we call weight W.

If ρ is constant, weight varies as shown by the straight line in Figure A7 – 1 which shows that weight decreases both above and below the Earth's surface.

Funny Facts of Physics

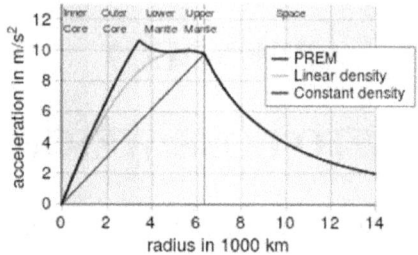

Figure A7–1. Weight Inside and Outside of the Earth

The density of the Earth is not constant. If you assume it is most dense at the center of the Earth and it gets less dense linearly with the distance from the center to the surface, Figure A7 – 1 would change slightly. This is shown as the line above the straight line in Figure A7 – 1. It is known, however, that the density of the Earth is more complicated than the densities assumed above. In the center of the Earth is a mass of iron, then there are other layers as shown in Figure A7 – 2.

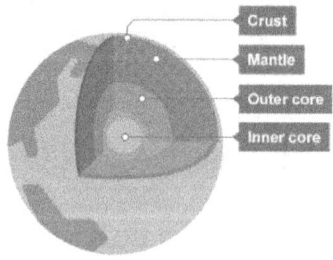

Figure A7 – 2. Composition of the Earth

The crust of the Earth is about 5–70 km deep. The mantle goes to a depth of about 2,890 km. The core is in two parts. The inner core is solid with a radius of about 1,220 km. The outer core is a liquid extending to a radius of about 3,400 km. The densities

How To Lose Weight

are between 9,900 and 12,200 kg/m^3 in the outer core and 12,600–13,000 kg/m^3 in the inner core. Assuming these density variations one gets the top line in Figure A7 – 1.

Even taking these internal features of the Earth into account the weight of an object would not be much different from that using the assumption of a constant density. The qualitative result that weight would be lost as one went deeper inside the Earth would be the same as if the density of the Earth was constant.

Funny Facts of Physics

APPENDIX 8

THE MOON IS FALLING

The relationship between height and time of fall is governed by the equation
$$h = (½) g t^2$$
where h is the height above the Earth and t is the time it takes to fall to the Earth. From this equation, the time for an object to fall is given by the equation
$$t = (2 h/g)^{½}$$
where ½ means the square root.

The distance x it will move horizontally is given by
$$x = v t$$
and using t from the equation above,
$$x = v (2 h/g)^{½}.$$

For an object to stay in orbit around the Earth its centripetal force must equal the gravitational force. In equation form that is
$$mv^2/r = GMm/r^2$$
where m is the mass of the object, v is its speed, r is its distance from the center of the Earth, G is Newton's Gravitational constant = 6.67 x 10^{-11} m^3/kg s^2, M is the mass of the Earth = 5.98 x 10^{24} kg, and the radius of the Earth r = 6.36 x 10^6 m. So the speed is given by
$$v = (G M/ [r + h])^{½}$$
where h is the height above the surface of the Earth, and ½ is the square root.

Funny Facts of Physics

APPENDIX 9

FOR A GOOD STRETCH TRY A BLACK HOLE

For decades investigators of black holes were convinced that everything that fell into a black hole was doomed to be there forever. Nothing could escape. That included even light which has no mass.

In the 1960s John Wheeler, a physicist led a team that stated that black holes "have no hair." They meant that black holes were identified by their spin, angular momentum, and mass and that nothing extended from them.

A decade later Stephen Hawking proposed that black holes emit quantum particles called Hawking radiation. So he proposed that black holes do have hair. If this were true then eventually, Hawking radiation would cause black holes to completely evaporate.

Funny Facts of Physics

APPENDIX 10

THE FASTEST

In general an equation for this addition of speeds is
$$w = u + v$$
in which, u = speed of the boat in the river or person on the train, v = speed of the river or train and w = speed of the boat in the river or the person on the train as observed by someone standing still.

For speeds close to that of light. if a source of light moves at speed u and emits a light at speed c then the speed of light relative to another observer is, according to the special theory of relativity,
$$w = (u + v)/[1 + (u\ v/c^2)].$$
That equation is like the one about the river or train except for the extra term $[1 + (uv/c^2)]$.

For the case of a plane moving at 500 mph with a tailwind of 50 mph, using Einstein's equation we get
$$w = 550/\ [1 + (500/45 \times 10^{16})$$
$$= 550/[1 + (11 \times 10^{-16})]$$
which is for all practical purposes indistinguishable from 550 mph.

Funny Facts of Physics

APPENDIX 12

SHRINK HAPPENS

The equation given by Einstein's Special Theory of Relativity for the contraction in length of a moving object is given by

$$L' = L/\sqrt{[1-(v^2/c^2)]}$$

where L' is the length of the object seen by a stationary observer, L is the length seen by an observer on the moving object, v is the constant speed of the moving object, and c is the speed of light. Note that if v is very small compared to c then L' ≈ L.

Funny Facts of Physics

APPENDIX 13

HOW TO STAY YOUNG

The equation given by Einstein's Special Theory of Relativity for the dilation of time Δt of a moving object compared to the time of a stationary observer Δt_s is given by

$$\Delta t = \gamma \, \Delta t_s = \Delta t_s / [1 - (v/c)^2]^{1/2}$$

which can be written

$$\Delta t_s = \Delta t \, [1 - (v/c)^2]^{1/2} = \Delta t / \gamma.$$

In this equation v is the constant speed of the moving object, c is the speed of light, and ½ means square root.

Funny Facts of Physics

APPENDIX 14

HOW TO BULK UP

The equation for increase of mass with speed given by Einstein is
$$m = m_0 / [(1-v^2/c^2)]^{1/2} = \gamma\, m_0$$
where m is the mass of an object moving at speed v, m_0 is its mass at rest, c is the speed of light, and ½ means square root. The quantity γ is reproduced as it is in Chapters 12 and 13 in Figure 12 – 4.

Funny Facts of Physics

APPENDIX 16

SCHRÖDINGER'S PATCHWORK

Schrödinger's equation is expressed as

$$- (h^2/8 \, m \, \pi^2) \, \nabla^2 \, \Psi + V \, \Psi = - (h/2 \, \pi \, I) \, \partial \Psi / \partial t$$

where the solution Ψ is dimensionless, h is Planck's Constant, and m is the mass of the particle. The term $\partial \Psi / \partial t$ is the derivative of the solution with respect to time. The derivative means a change in Ψ as time changes. The term ∇^2 is the second derivative with respect to x, y, and z. That means the change in the change of Ψ as the position of the particle changes. The quantity Ψ is a function of z, y, z, and is called a wave function. The three terms in the equation have the dimension of energy so Ψ is called the energy wave function.

The related time independent equation Schrödinger constructed by replacing the last term by the energy of the particle multiplied by the wave function is

$$- (h^2/8 \, m \, \pi^2) \, \nabla^2 \, \Psi' + V \, \Psi' = E \, \Psi'.$$

The solutions to this equation give the discrete (quantum) energy levels the particle can have. They are called eigenvalues.

Funny Facts of Physics

APPENDIX 17

YOU CAN NEVER BE CERTAIN

When making a simultaneous measurement of the position and momentum of an atomic size (or smaller) particle, there is an inherent uncertainty in the values of the measurements given by the Heisenberg Uncertainty Principle. Written as an equation it is
$$\Delta x \, \Delta p \geq h/4\pi$$
where x = position of the particle, p = mv = the momentum of the particle, m is its mass, v is its velocity, and h = Planck's Constant = 6.63×10^{-34} Joule-sec. The symbol Δ means a range of values or the uncertainty in a measurement.

The more accurately you know x (a smaller Δx) the less accurately you can know p, (a larger Δp) and vice versa.

Consider a macroscopic case of a baseball. For a thrown baseball of mass = 0.145 kg assume its speed is v = 100 miles per hour = 44.7 m/s. Then p = 0.145 kilograms x 44.7 m/s = 6.48 kg-m/s. The mound is x = 90 feet from the plate but the ball can leave the pitcher's hand as much as 2 feet in front of the rubber. The catcher may be as much as 2 feet behind the plate. Then you can say the distance the ball will go is 90 feet plus or minus 4 feet. Δx = 4 feet = 1.22 m. Then Δp would be

$$\Delta p > (h/4\pi)/ \Delta x = 5 \times 10^{-35}/1.22 = 4.1 \times 10^{-35} \text{ kg-m/s}$$

which gives

$$\Delta v = 4.1 \times 10^{-35}/ 0.145 \text{ kgm} = 2.8 \times 10^{-34} \text{ m/s}$$

which is an insignificantly small number.

This means that though the location of the baseball in the catcher's glove has an uncertainty of 4 feet the speed of the ball

Funny Facts of Physics

measured as 100 miles per hour is exact to within a negligible uncertainty in its velocity.

Now consider a proton of mass m = 1.67×10^{-27} kg moving at one half the speed of light ($0.5c = 1.5 \times 10^8$ m/s). You can measure its position with a detector. For example say it has an uncertainty in this measurement of where it is at the moment of measurement of 10^{-15} m. That is its classical radius of 10^{-15} m. Then a simultaneous measurement of its momentum will have an uncertainty of

$$\Delta p \geq (h/4\pi)/\Delta x = 5.27 \times 10^{-35}/10^{-15}$$
$$= 5.27 \times 10^{-20} \text{ kg-m/s.}$$

The error in measuring its velocity would be

$$\Delta v \geq \Delta p/m = 5.27 \times 10^{-20}/1.67 \times 10^{-27}$$
$$= 3.16 \times 10^7 \text{ m/s}$$

which gives the uncertainty as a fraction of its speed $\Delta v/v = 0.2$. Thus a measurement of its speed would be accurate only to about 20% of its speed

APPENDIX 19

BATHED IN PHOTONS

The electromagnetic spectrum is a band of waves of different energies from gamma rays to radio waves as shown in Figure 1–1.

An electromagnetic wave, as its name implies, is made of the combination of an oscillating electric field E and an oscillating magnetic field B. The energy per unit volume, W, associated with the various components of the electromagnetic spectrum at different frequencies is given by a combination of E^2 and B^2, specifically

$$W = (\varepsilon_0 E^2 /2) + (B^2/2\mu_0)$$

where ε_0 is the permittivity of free space and μ_0 is the permeability of free space.

Permittivity is a measure of how much an electric field affects a medium, and permeability is the ability of a medium to support a magnetic field.

Some examples of these quantities relative to their values in free space are given in Table A19 – 1.

Material	(ε_r)	(μ_r)
Air	1	1
Water	50	1
Glass	10	5
Quartz	5	Insulator
Steel	Conductor	280000

Table A19 – 1. Relative Permittivity and Permeability

From Table A19 – 1 we see that materials like steel, copper, aluminum, for example, with high permittivity are good

Funny Facts of Physics

conductors of electricity. Those materials with low permittivity are good electrical insulators. Also from Table A19 – 1 we see that materials like steel with high permeability make efficient magnets.

When electromagnetic energy is thought of not as waves but as a train of photons the energy, E, of each photon is

$$E = hf = hc/\lambda$$

where $h = 6.63 \times 10^{-34}$ Joule-sec is called Planck's Constant. The quantities f and λ are the frequency and wavelength of an electromagnetic wave, and c is the speed of light.

Electromagnetic radiation considered either as photons or waves have no mass but have momentum, p given by

$$p = hf/c.$$

APPENDIX 21

COLDER THAN THE COLDEST

Familiar Temperature Scales
Two equations that relate the Celsius and Fahrenheit temperature scales are
$$C = (F - 32) \, 5/9$$
$$F = (9/5) \, C + 32.$$

For example if you read 70 on the Fahrenheit scale you subtract 32 to get that scale even with the Celsius scale. Then multiply by $100/180 = 5/9$. You get $(70 - 32) \times 5/9 = 21$ C. If you started with 15 Celsius multiply by 9 and divide by 5. Then add 32. You get $(15 \times 9/5) + 32 = 59$ F.

Absolute Temperature Scales
Temperature is a measure of the speeds at which molecules or atoms in a substance are moving. The faster they move the higher the temperature. The slower they move the lower the temperature. If they stop moving, the temperature should be at its lowest. Call that the lowest Absolute Temperature and set it at zero degrees. The Absolute Temperature scales, Kelvin and Rankine, are not based on the freezing and boiling points of water but on the triple point of water (the temperature and pressure at which water coexists in the solid, liquid, and gaseous phases). Equations that relate the Kelvin and Rankine scales to the Celsius and Fahrenheit scales are
$$K = C + 273.15$$
$$R = F + 459.69.$$

This means that the K = 0 Absolute Temperature is -273.15 on the C scale, and on the R Absolute Temperate scale R =0 gives F = -459.69. No conventional temperature can go below absolute zero.

Funny Facts of Physics

Energy Levels

In a population of quantum particles, say atoms, they arrange themselves in different higher energy levels. For example, a two level system would look like Figure A21 – 1.

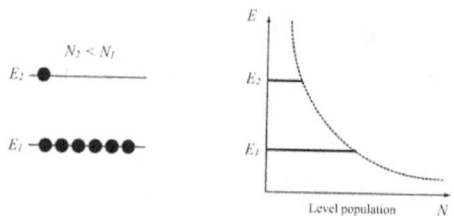

Figure A 21 – 1. Energy Level Diagram

In Figure A21 – 1 the lengths of the horizontal lines represent the number of particles with the energies associated with those line. The mathematical model that describes the normal distribution of energy levels in a population of atoms is
$$N_H / N_0 = e^{-E_H / kT}$$
where e = 2.71828. . . . is the exponential, E_H is the energy of the higher energy level given by H, N_H is the number of atoms in the higher energy level H, N_0 is the number of atoms in the ground energy level, k = 1.38 x 10^{-23} Joule/Kelvin is the Boltzmann constant, T is the Absolute Temperature in Kelvin degrees in this case. For normal situations N_0 is larger than N_H.

Taking the natural logarithm of both sides of the equation above gives
$$- E / kT = \ln (N_H / N_0) = \ln N_H - \ln N_0$$
from which one gets
$$\ln N_0 - \ln N_H = E / kT$$

Population Inversion

If a population of atoms is so highly irradiated at the energy that excites ground state atoms to a higher state fast enough, and they decay slow enough, there could be more atoms

Colder Than The Coldest

in the higher energy state than in the lower energy state. This is called a population inversion and is shown in Figure A21 – 2.

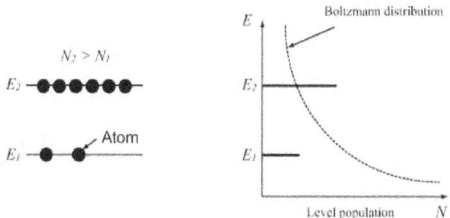

Figure A 21 – 2. Inverted Energy Level Diagram

An inverted population is one with N_0 smaller than N_H which, in the equation above, makes $\ln N_H$ larger than $\ln N_0$ so E / kT in the equation above must be a negative quantity. But E and k are positive quantities thus T must be negative.

Funny Facts of Physics

APPENDIX 22

THE VIRTUAL VACUUM

Black Body

A Black Body is any object that is a perfect emitter and a perfect absorber of radiation, is at a uniform temperature, and whose radiation depends only on its temperature. Black-body radiation is the thermal electromagnetic radiation emitted by a Black Body. Bodies that are not quite black are called Gray Bodies. Most times the laws of black bodies are good approximations for gray bodies.

Black Body Power

The radiation from a Black Body (BB) has a specific spectrum and intensity distribution that depends only on the body's temperature. The power of the radiation depends on the fourth power of the temperature as in the equation

$$P = \sigma A \varepsilon (T^4 - T_0^4)$$

where T is the absolute Kelvin temperature of the body, T_0 is the ambient absolute Kelvin temperature, ε is the emissivity of the body which varies for different materials, A is the surface area of the body, and σ is the Stefan-Boltzman constant = 5.67×10^{-8} W/m^2 K^4.

The radiation emitted by many objects can be approximated as black-body radiation. A good example of a Black Body is say a tennis ball at some temperature. The radiation inside is Black Body radiation. That radiation can be sampled by making a small hole and examining the radiation that emerges.

Ambient temperature is about 300 K. Human bodies are at about 100 K. Those two temperatures used in the equation for power form the basis for devices such as infrared night vision goggles.

Wavelength Spectrum

The distribution of the energy emitted per unit volume per unit wavelength at different wavelengths of the radiation for a specific temperature is given by the Planck equation

Funny Facts of Physics

$$I(\lambda, T) = (2 h c^2 / \lambda^5) / [\exp(h c/\lambda k T) - 1]$$

where I = irradiance in W/m² -sr -μm, λ = wavelength of radiation, T = Kelvin temperature, h = Planck's constant, c = speed of light, k = Boltzmann's constant, and exp stands for the exponential e. The Plamck equation is plotted against wavelength (λ) in Figure A22 – 1 for various temperatures. Ambient temperature is about 300 K.

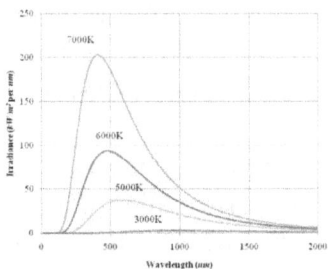

Figure A22 – 1. B B Irradiance vs. λ and T

Cosmic Background Radiation

The CBR is essentially Black Body radiation from the universe that existed at the Big Bang and has cooled to a temperature of 2.7 K. It is plotted in Figure A22 – 2.

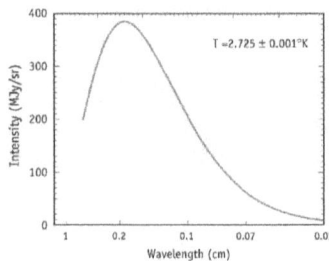

Figure A22 – 2. C B R vs. λ

Figure A22 – 2 compares well with that of the theoretical black body spectra in Figure A22 – 1 which confirms that the CBR is Black Body radiation.

APPENDIX 24

TIME WILL TELL

A fundamental dimension of nature, length, is given by a combination of three fundamental constants of nature: Planck's constant h = 6.63 x 10^{-34} Joule-sec (named after the German physicist Max Planck who founded quantum mechanics), the speed of light c = 3 x 10^8 m/sec from Einstein's theory of relativity, and Newton's universal gravitational constant G = 6.67 x 10^{-11} Newton-m^2/kg^2. When the constants are arranged as $(Gh/c^3)^{1/2}$ where ½ means square root. This grouping has the dimension length and is called the Planck length. It has the value 4.05 x 10^{-35} meters. That is a very small length.

Planck time is equal to the Planck length divided by the speed of light. It is given by the equation

$$t_P = [hG/c^5]^{1/2} = 1.35 \times 10^{-43} \text{ sec.}$$

That is a very short time.

Some definitions of Planck length and Planck time use h/2π instead of h thus have slightly different values (by a factor of 2π) for the Planck length and Planck time.

At the Planck time, 1.35 x 10^{-43} sec, the density of the universe is thought to have been about 10^{93} g/cc. An object of such small size with that mass would be like a quantum black hole. The known laws of physics might not apply at these dimensions. A yet to be developed quantum gravity theory or its possible alternate or successor may have to be used to explain phenomena under those conditions.

Funny Facts of Physics

INDEX

Absorption. 39-41
Andromeda galaxy. 11, 115
Anthropic cosmological principle. 23, 153
Arrow of time. 52-55
Atom bomb. 46, 50
Big bang. 54, 123, 124, 129, 134-136, 139-143, 148, 162, 212
Black body. 123, 211, 212
Black hole. 7, 8, 69-71, 129, 162, 191, 213
Bohr. 108, 152
Boltzmann. 208
Born. 123, 129, 138, 141, 158
Cerenkov radiation. 11, 77, 78
Composition of the Earth. 11, 186
Conservation of energy. 45, 99
Cosmic background radiation. 11, 123, 212
Dirac. 152
Einstein. 49, 76, 83, 85, 86, 93, 99, 100, 108, 144, 148, 149,
152, 161, 163, 166, 199
Electromagnetic spectrum. 11, 25, 26, 31, 42, 77, 112, 115, 205
Emission. 42, 43, 77, 93, 134
Energy ladder. 11, 46, 47, 49, 52
Energy level. 11, 121, 122, 208, 209
Feynman. 100, 152, 153
Fine structure constant. 153
Gell Mann. 152
Gravitational waves. 11, 148, 149, 152, 166
Gravity. 4, 7, 19, 57, 59-61, 63, 65, 67, 69-71, 129, 213
Hawking. 158, 159, 191
Heisenberg. 124, 152, 203
Infrared. 26, 39, 41, 42, 112, 113, 128, 211
Lee. 152
Length contraction. 11, 13, 81-84, 89
LIGO. 149
Mechanical energy. 46, 49, 52
Metaphysics. 4, 151, 169, 171, 172
Michelson. 157
Milky way. 28, 79, 115, 117, 128, 143

INDEX (Cont'd.)

Moon. 7, 8, 11, 28, 41, 63, 66-69, 161, 189
Negative temperature. 121
Newton. 66, 161, 185, 213
Penumbra. 11, 41
Permeability. 13, 205, 206
Permittivity. 13, 205, 206
Photoelectric effect. 93, 95
PIN Code. 8, 151, 154
Pinhole camera. 93, 94
Planck. 135, 212, 213
Podolsky. 108
Pseudophysics. 4, 169, 171-173, 175, 177
Reflection. 39-41
Renewable energy. 43, 45, 50
Rosen. 108
Satellites. 31, 87, 112, 116, 167
Scattering. 39, 41
Schrödinger. 99-101, 108, 201
Slit. 11, 94-97, 100, 176
Solar. 28, 43, 45-47, 116, 167
Spaghettification. 11, 70
String. 67, 163, 164
Theory of everything. 158-164, 166
Time dilation. 13, 86, 87, 89
Transmission. 39, 41
Twin paradox. 87
Umbra. 11, 41
Uncertainty principle. 104, 105, 124, 203
Virtual vacuum. 8, 9, 123, 148, 211
Wave function. 97, 99, 101, 109, 158, 201
Wind. 43, 45-47, 75, 76, 116
Yang. 152

Funny Facts of Physics

ABOUT THE AUTHOR

Dr. Veigele earned a B. A. in mathematics and an M. A. in physics from Hofstra College, New York and a Ph. D. in physics from the University of Colorado, Boulder, Colorado. He taught and did research at Williams College, Massachusetts; Hofstra College, New York; University of Colorado, Boulder, Colorado; and University of California, Santa Barbara, California. He was a Professor, Head of a Physics Department, and a research scientist. He is listed in many Who's Who publications and has received honorary research and publication awards. Dr. Veigele continues to be a frequently invited speaker on science and environmental issues.

The author has published more than seventy articles and stories in technical journals and magazines and has published thirteen books

www.ingramcontent.com/pod-product-compliance
Lightning Source LLC
Chambersburg PA
CBHW050207230526
45470CB00001B/279